Why We are Free

Why We are Free

Consciousness, Free Will and Creativity in A Unified Scientific Worldview

DAVID LAYZER

Author of *Cosmogenesis:*
The Growth of Order in the Universe
First Donald H. Menzel
Professor of Astrophysics
Harvard University

Anthony Aguirre and Bob Doyle, Editors

Publisher's Cataloging-in-Publication Data
(Prepared by The Donohue Group, Inc.)

Names: Layzer, David, 1925-2019, author. | Aguirre, Anthony, 1973- editor. | Doyle, Bob, 1936- editor.

Title: Why we are free : consciousness, free will and creativity in a unified scientific worldview / David Layzer, first Donald H. Menzel Professor of Astrophysics, Harvard University ; Anthony Aguirre, UC San Diego and Bob Doyle, Harvard University, editors.

Description: First edition. | Cambridge, MA, USA : I-Phi Press, 2021. | Includes bibliographical references and index.

Identifiers: ISBN 9780983580256 (paperback) ASIN B08XNZR3FF (Kindle)

Subjects: LCSH: Free will and determinism. | Cosmology. | Philosophy and science.

Classification: LCC BJ1461 .L39 2021 | DDC 123.5--dc23
1. Free Will. 2. Cosmology. 3.Philosophy. I. Title.

I-Phi Press
Cambridge, MA, USA

For my undergraduate and graduate students,
from whom, over many decades,
I learnt more than they learned from me

Preface

David Layzer was an astrophysicist and cosmologist at Harvard who from the 1960s onward generated key insights into the evolution and structure of the universe, many of which arguably remain under-appreciated. He was among the first to carefully and incisively analyze the arrow of time in cosmology, he formulated a subtle, distinct, and fascinating version of the "cosmological principle," and he provided key insights into the growth of both *entropy* and *order* through cosmic evolution. In his later career he focused on the connection between cosmic foundations of randomness, the growth of order, and the creativity and freedom inherent in biological evolution and mental processes. *Why We are Free* ties all of these threads together. The work speaks for itself, but readers may find some historical and scientific contextualization of each of them useful.

Layzer's cosmological work began at a time when our view of cosmic history was far less settled. In the early 1960's he was invited to a conference on the nature of time at Cornell University organized by the founders of the steady-state theory of the universe, Hermann Bondi, Tommy Gold, and Fred Hoyle.

The steady-state founders went beyond Newton's "cosmological principle" that space is homogeneous and isotropic, the same in all places and in all directions. They added their "perfect cosmological principle," claiming the universe is eternal and appears the same at all times. This appeared to eliminate what Arthur Stanley Eddington in 1928 called "Time's Arrow," which Eddington had associated with the apparently *irreversible* increase in entropy demanded by the second law of thermodynamics.

At the Cornell conference Layzer proposed his "strong cosmological principle," that the homogeneity and isotropy of matter and space applies both statistically and exactly. The subtle but foundational implication of this assumption is that the cosmological principle is not just a convenient approximation to a more precisely knowable cosmic state, but that the statistical description is complete and cannot be improved upon. This enormously reduces, for example, the information required to specify the initial cosmic state.

Subsequently, Layzer argued that the strong cosmological principle implies a fundamental uncertainty inherent in any finite physical system: because the statistical description of the universe is complete, the precision with which any subsystem can be described is limited. He called this implication "primordial randomness." He later came to believe that this very uncertainty is at the root of quantum uncertainty. Sketched in his book *Cosmogenesis*, this idea was revisited by Layzer[1], and by Max Tegmark and one of us[2], eventually splitting into two versions of a "Cosmological interpretation of quantum mechanics," similar in spirit but different in details and context.

Layzer also showed in detail different types of arrows of time, and some of their relations: alongside Eddington's identification of increasing entropy as the fundamental time's arrow, Layzer now added another arrow he called the "historical arrow of time; this was in addition to the so-called "radiation arrow" or "electromagnetic arrow" (the asymmetry between emission and absorption processes) and the so-called "cosmological arrow" (the expansion of the universe).

The strand of Layzer's work addressing the "arrow of time" also began early in the origins of modern cosmology. The two most important publications were in *Scientific American* in 1975 and the *Astrophysical Journal* in 1976. In the former article, Layzer introduced his famous example of the perfume bottle being opened and the perfume dispersing into the room.

1   See https://arxiv.org/abs/1008.1229 for Layzer's elegant and detailed exposition.
2   Aguirre, Anthony, and Max Tegmark. "Born in an infinite universe: a cosmological interpretation of quantum mechanics." Physical Review D 84.10 (2011): 105002.

Layzer likened the time-reversed process to a movie played backwards. If the "initial" state contained "hidden" microscopic information corresponding to a reversal of the momentum of each perfume molecule at a certain time, classical dynamics says that the perfume will make its way back into the bottle. This is a powerful visual image illustrating Josef Loschmidt's reversibility objection to Ludwig Boltzmann's H-Theorem.

In Layzer's view, primordial randomness precludes just the sort of "hidden information" that would allow such a reversal.

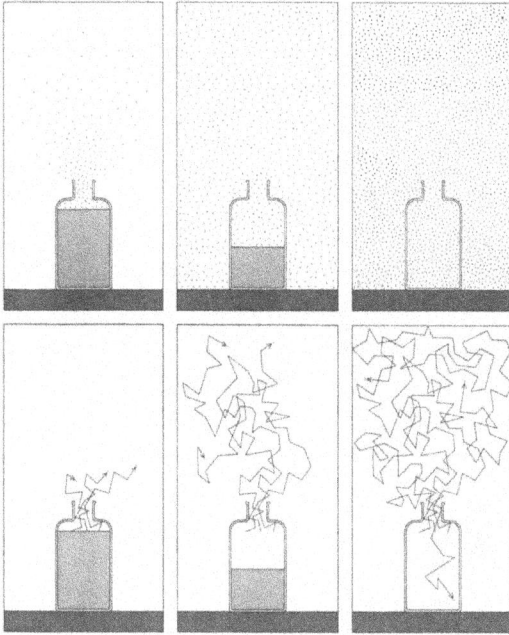

Source: "The Arrow of Time" *Scientific American*, December, p.57 (1975)

Beyond its clarity, what most distinguished Layzer's thinking about time's arrow in cosmology was his resolution of a very basic paradox: the universe appears to start in a dense, high-energy, equilibrium state. Yet the second thermodynamic law implies that entropy increases. So how does the universe ever leave equilibrium and so generate all of the ordered and highly non-equilibrium structures it clearly contains? Layzer's resolution, in his words from the 1975 article:

> Suppose that at some early moment local thermodynamic equilibrium prevailed in the universe. The entropy of any region would then be as large as possible for the prevailing values of the mean temperature and density. As the universe expanded from that hypothetical state the local values of the mean density and temperature would change, and so would the entropy of the region. For the entropy to remain at its maximum value (and thus for equilibrium to be maintained) the distribution of energies allotted to matter and to radiation must change, and so must the concentrations of the various kinds of particles. The physical processes that mediate

these changes proceed at finite rates; if these "equilibration" rates are all much greater than the rate of cosmic expansion, approximate local thermodynamic equilibrium will be maintained; if they are not, the expansion will give rise to significant local departures from equilibrium. These departures represent macroscopic information; the quantity of macroscopic information generated by the expansion is the difference between the actual value of the entropy and the theoretical maximum entropy at the mean temperature and density.[3]

This core and clear insight — that information is the gap between realized and possible entropy, and that the latter increases in an expanding universe — has even now not fully penetrated the thinking of many working in the field. Going further, Layzer persuasively argues that this dynamic, in which would-be equilibrating interactions fail to maintain equilibrium, pervades and underlies the growth of order in the universe. For Layzer, this along with the

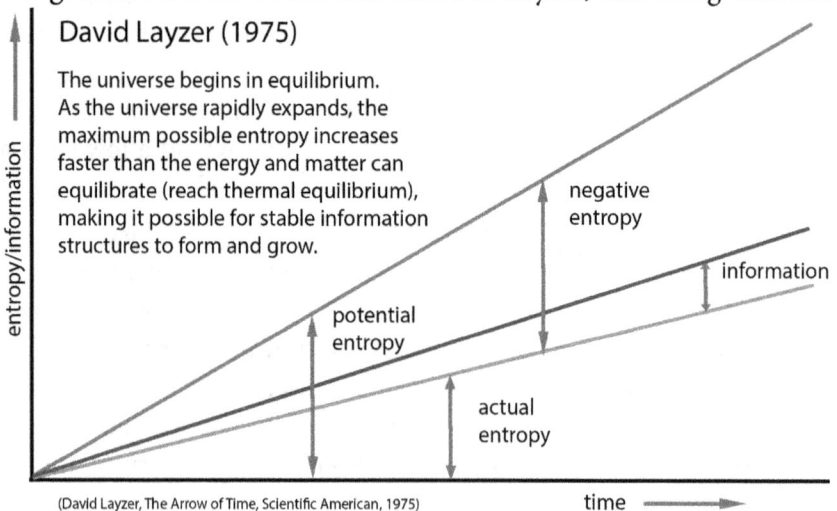

David Layzer (1975)

The universe begins in equilibrium. As the universe rapidly expands, the maximum possible entropy increases faster than the energy and matter can equilibrate (reach thermal equilibrium), making it possible for stable information structures to form and grow.

(David Layzer, The Arrow of Time, Scientific American, 1975)

strong cosmological principle provided a clear explanation of the arrows of time, with the cosmological principle ensuring statistical equilibrium at the earliest time, and the time-varying rates of various processes during cosmic evolution doing the rest. Again in his words:

> As a result the big bang is an exceedingly gentle process; local equilibration processes easily keep pace with the changing macroscopic conditions of temperature and density during the first fraction of a microsecond. It is only for this brief initial phase in

3   "The Arrow of Time" *Scientific American*, December, p.68 (1975).

the evolution of the universe that local thermodynamic equilibrium can be assumed, but from that assumption it follows that the expansion of the universe has generated both macroscopic information and entropy. Thus the cosmological arrow, the historical arrow and the thermodynamic arrow all emerge as consequences of the strong cosmological principle and the assumption that local thermodynamic equilibrium prevailed at or near the initial singularity. Remarkably, neither of these assumptions refers directly to time or temporal processes.[4]

For Layzer, these cosmological considerations tied directly to basic issues about the nature of complex systems including biological ones. He explored these ties in detail in his 1990 book, *Cosmogenesis*, which lays out a full vision for how the enormous amount and quality of structure comes into being, first through cosmic processes and then via life and evolution that can "utilize" cosmic and astrophysical order. That book also began Layzer's study of free will, which through his arguments he connected to randomness in the earliest time.

In his later years, Layzer developed through a series of talks and unpublished papers a compelling view of how the freedom and creativity of living and mental systems can coexist with the (seemingly) rigid determinism of natural law. It has been argued in many places that randomness — quantum or otherwise — does not bear upon free will. Layzer forcefully disputes this, arguing that the same randomness and growth of order that pervades the universe underlies living phenomena, with the randomness providing the fuel (but not in itself an explanation) for genuine novelty and choice.

In *Why We Are Free*, his last work (especially chapter VII and the following four sections), we see Layzer 's concise telling of this entire story.

Anthony Aguirre, UC Santa Cruz

Bob Doyle, Harvard University

January, 2021

---

4   "The Arrow of Time" *Scientific American*, December 1975, p.57.

Contents

What This Book Is About

No biological capacity is more distinctively human than our ability to shape future events in ways that accord with our needs, wants, and desires. Yet many mainstream scientists and philosophers deny the existence of such a capacity. They subscribe to a scientific worldview called physicalism, or materialism, according to which reality comprises just those objects and relations that figure in our strongly confirmed physical laws. Francis Crick, co-discoverer with James Watson of DNA's double helix, summarized this idea in the opening paragraph of his book *The Astonishing Hypothesis, the Scientific Search for the Soul:*. It is the alien idea that each person is no more than a vast assembly of nerve cells and their molecules, that we are "nothing but a pack of neurons."

This book describes an alternative to physicalism in which our joys and our sorrows, our memories and our ambitions, our sense of personal identity and free will, are just as real as the objects and relations of the world physics describes. Like physicalism, the proposed worldview rests on our strongly confirmed physical laws. I argue that these laws don't need to be supplemented by additional laws that expand the domain of natural science to include consciousness and subjectivity. Physicalism, I argue, rests on Newton's assumption in The Principia that place is absolute. This assumption implies, for instance, that the present value of the mass per unit volume at the Sun's position has a definite value (though one we can't know with infinite precision).

This book proposes to replace the assumption that place is absolute by a pair of less obvious cosmological assumptions: The first of these, the strong cosmological principle, says that a complete description of the Universe doesn't privilege any position or direction in space. It implies that a complete description of the Universe wouldn't tell us the values of macroscopic quantities. Instead it would specify probability distributions of the possible values of macroscopic quantities. It would interpret the probability of any given range of possible values of a macroscopic quantity as the relative frequency of that range in an infinitely extended sample of the Universe. The second assumption, the assumption of primordial randomness, says that the probability distributions that characterize the Universe's initial state were maximally random. Randomness in this context is synonymous with statistical entropy, a property of probability distributions introduced in 1872 by Ludwig Boltzmann as the counterpart in the atomic theory of gases to Rudolf Clausius's thermodynamic entropy. The assumption of primordial randomness implies that the Universe has evolved from a state of complete disorder, so it contradicts the widely believed claim that the entropy of the Universe never decreases. This book argue that this generalization of the strongly confirmed law of entropy non-decrease for undisturbed macroscopic systems is untenable.

Chance and the openness of the future are pervasive features of prescientific views of the natural world. Physicalism tells us that they are illusory. In a scientific worldview that incorporates the strong cosmological principle and the assumption of primordial randomness chance is indeed pervasive and the future is indeed largely open. But how do physics and biology themselves fare under the new scientific dispensation?

I will argue that the two cosmological hypotheses allow a more unified view of macrophysics, microphysics, and statistical physics (which links macrophysics to microphysics). The implications of the two cosmological hypotheses for biology are more striking. Physicalism purports to reduce biology to physics. But physicalism attributes macroscopic chance to human ignorance. By contrast, macroscopic chance plays a central role in evolution, both in genetic variation (for example in genetic recombination) and natural selection (think of extinction events). Consequently, although biological processes are governed by physical laws, biology doesn't reduce to physics. I will argue that the strong cosmological principle and the assumption of primordial randomness together with our strongly confirmed physical laws provide a framework for modern evolutionary theory. They also provide a framework for understanding consciousness, creativity, and free will as biological phenomena.

I
Physics, Biology, and Physicalism

A physical theory needn't be intuitively plausible, but it must make testable predictions that agree with experiment or observation. For example, quantum mechanics describes electrons as mathematical points endowed with mass and electric charge. It assigns these points spatial coordinates and asserts that when an electron is in a definite physical state its coordinates don't have definite values. None of this makes intuitive sense. How could a mathematical point have a finite mass? How could a point's coordinates not have definite values? Nevertheless, unambiguous rules link quantum mechanics' abstract mathematical laws to the outcomes of a wide range of experiments, and the close agreement between the observed and predicted outcomes of these experiments leaves no room for doubt that the laws are, at least, very nearly correct. Experience isn't just the arbiter of scientific theories, though. It is itself a subject of scientific investigation. Neuroscientists investigate how perception, memory, cognition, and other mental states and processes are linked to physical structures and processes in the brain. Such studies have produced convincing evidence that the mathematical laws that govern physical phenomena also govern the biological processes that accompany mental processes.

But mental processes seem to be more than the physical processes that accompany them. All experience is someone's experience. My experience of seeing a red ball is distinct from your experience of seeing what we both believe to be the same ball. It may seem obvious that this aspect of experience – consciousness – is as much a part of the natural world as matter and energy. Yet there is no scientific or philosophical consensus about how – or even whether – consciousness fits into a scientific picture of the world.

Some scientists and philosophers argue that the subjective aspect of consciousness – the aspect that makes your consciousness different from mine – isn't part of a scientific picture of the world. Francis Crick, co-discoverer with James Watson of DNA's double helix, summarized this position in the opening paragraph of his book *The Astonishing Hypothesis, the Scientific Search for the Soul*:

> The Astonishing Hypothesis is that "You," your joys and your sorrows, your memories and your ambitions, your sense of personal identity and free will, are in fact no more than the behavior of a vast assembly of nerve cells and their associated molecules. As Lewis Carroll's Alice might have phrased it: "You're nothing but a pack of neurons." This hypothesis is so alien to the ideas of most people alive today that it can truly be called astonishing.[1]

Crick's astonishing hypothesis implies that biology not only rests on physics – something few if any contemporary biologists would dispute – but also reduces to physics: the natural world contains just those entities mentioned in our strongly confirmed physical theories, nothing more. I'll refer to this scientific worldview as *physicalism*.

Among the features of conscious life that physicalism reduces to the behavior of assemblies of brain cells is free will: our felt ability to alter the course of events through our deliberate acts. Though habits and deep-seated preferences dictate many of the choices we make in our daily lives, we take it for granted that the actions that flow from decisions we have thought about long and hard help shape the future. We think the world would have been different in ways that matter to us if we had decided and acted differently. Law, ethics, and widely held opinions about the aims of education all presuppose that our deliberate actions spring from a kind of freedom other animals don't enjoy. Our laws punish theft, but we don't regard a dog that steals another dog's bone as a lawbreaker. We think torture is wrong, but we don't think a cat is acting unethically when it tortures a mouse. We train our pets, but we believe children should

---

1   Crick, Francis. *The Astonishing Hypothesis* (New York: Simon and Schuster 1994) p. 3.

be brought up not just to behave in certain ways but also to make good decisions. None of these beliefs and attitudes would be tenable if we weren't capable of making choices and decisions that affect the future course of events. Yet decision-making is a biological process, and biological processes are also physical processes, governed by the physical laws that prevail in physics and chemistry laboratories and in stars and galaxies. These laws are deterministic in the following sense: They connect a complete description of the present physical state of an isolated, or undisturbed, physical system to a complete description of any of the system's past and future states. The laws of quantum physics are just as deterministic in this sense as those of classical, or pre-quantum, physical theories – theories that deal with macroscopic and astronomical phenomena. Quantum mechanics' law of change, like its classical counterparts, links the present state of an undisturbed physical system to each of the system's future states.

Unlike classical states, quantum states of undisturbed systems aren't directly observable. The standard formulation of quantum mechanics links them to measurement outcomes by a rule that Paul Dirac in *The Principles of Quantum Mechanics* called a "general assumption." Given a system's quantum state, this rule (discussed in more detail below) enables one to calculate not only the possible outcomes of a measurement of any of the system's physical properties, such as position or momentum or energy, but also the relative frequency of each possible outcome in a long series (or large collection) of identical measurements. This feature of quantum mechanics – the unpredictability of individual measurement outcomes, along with the predictability of possible measurement outcomes and their relative frequencies in long series of identical measurements – is what physicists mean by "quantum indeterminism."

The philosopher John Searle assumes, as do many or perhaps most physicists, that "all indeterminism in nature is quantum inde-terminism."[2] He then argues that "consciousness is a feature of nature that manifests indeterminism" and concludes that quantum indeter-minism must underpin free will. While I agree with Searle that free will requires "indeterminism in nature," I think the conclusion that quantum indeterminism underpins free will is untenable, for two reasons. First, as far as I know, nothing in neuroscience suggests that

---

2   Searle, John R. *Freedom and Neurobiology* (New York, Columbia University Press, 2004) p. 74.

the neural processes accompanying decision-making are, or resemble, quantum measurements. Second, quantum measurements, unlike free acts, never have novel and unpredictable outcomes; their possible outcomes are entirely predictable. Yet if we have the ability to help shape future events through our deliberate actions, as I believe we do, the future must in some ways be open. Either free will is an illusion or determinism is false.

Like Crick, the evolutionary biologist Edward O. Wilson has embraced the first alternative:

> The self, an actor in a perpetually changing drama, lacks full command of its own actions. It does not make decisions solely by conscious, purely rational choice. Much of the computation in decision-making is unconscious – strings dancing the puppet ego. Circuits and determining molecular processes exist outside conscious thought. They consolidate certain memories and delete others, bias connections and analogies, and reinforce the neurohormonal loops that regulate subsequent emotional response. Before the curtain is drawn and the play unfolds, the stage has already been partly set and much of the script written.
>
> The hidden preparation of mental activity gives the illusion of will. We make decisions for reasons we often sense only vaguely, and seldom if ever understand fully. Ignorance of this kind is conceived by the conscious mind as uncertainty to be resolved; hence freedom of choice is ensured. An omniscient mind with total commitment to reason and fixed goals would lack free will.

But if free will is illusory, why is the illusion so strong? Wilson explains:

> ...Confidence in free will is biologically adaptive. Without it the mind, imprisoned by fatalism, would slow and deteriorate. Thus in organismic time and space, in every operational sense that applies to the knowable self, the mind does have free will.[3]

Psychologist Daniel Wegner agrees with Wilson that free will is an illusion. He argues that the neural processes underlying voluntary actions have two distinct outcomes: the action itself and awareness of willing the action. Is this awareness the cause of the action? From a thorough examination of the evidence bearing on this question, Wegner concludes the answer is no:

---

3   Wilson, E.O. *Consilience*, (New York, Knopf: 1998) p. 130.

> It usually seems that we *consciously* will our voluntary actions, but this is an illusion. ... *Conscious* will arises from processes that are psychologically and anatomically distinct from the processes whereby mind creates action. (pages1-2) ... If a team of scientific *psychologists* ... somehow had access to all the information they could ever want, ... they *could* uncover the mechanisms that give rise to all your behaviors .... [4]

Behind this thought is the image of a complex physical system, such as the solar system, whose physical states are completely determined by physical laws and antecedent conditions. Indeed Wegner ends his book with a quotation from Einstein:

> If the moon, in the act of completing its eternal way around the earth were gifted with self-consciousness, it would feel thoroughly convinced that it was traveling its way of its own accord. ...So would a Being, endowed with higher insight and more perfect intelligence, watching man and his doings, smile about man's illusion that he was acting according to his own free will.[5]

The philosopher Thomas Nagel has argued that physicalism can't be reconciled with the existence of consciousness:

> The existence of consciousness seems to imply that the physical description of the universe, in spite of its richness and explanatory power, is only part of the truth, and that the natural order is far less austere than it would be if physics and chemistry accounted for everything. If we take this problem seriously, and follow out its implications, it threatens to unravel the entire naturalistic world picture. Yet it is very difficult to imagine viable alternatives."[6]

The subtitle of *Mind and Cosmos* states that "the materialist neo-Darwinian conception of nature is almost certainly false."

If evolutionary theory is a purely physical theory, then it might in principle provide the framework for a physical explanation of the appearance of behaviorally complex animal organisms with central nervous system. But subjective consciousness, if it is not reducible to something physical, would not be part of this story; it would be left completely unexplained by physical evolution – even if the physical

---

4   Wegner, D.M. *The Illusion of Conscious Will*. (Cambridge: MIT Press 2002) p. 29.

5   Einstein, A. 1995. Quoted in Home, D. and A. Robinson, "Einstein and Tagore: Man, nature and mysticism", *Journal of Consciousness Studies* 2, pp.167-169.

6   Nagel, Thomas, *Mind and Cosmos: Why the Materialist Neo-Darwinian Conception of Nature Is Almost Certainly False* (Cambridge, Belknap Press of Harvard University) p. 35

evolution of such organisms is in fact a causally necessary and sufficient reason for consciousness.

Yet Nagel doesn't argue that consciousness lies outside the scope
of natural science. Rather, he suggests, "[i]t makes sense to seek an
expanded form of understanding that includes the mental but that is
still scientific – i.e. still a theory of the immanent order of nature."[7]

Ernst Mayr, an architect of contemporary evolutionary theory,
would have agreed. He wrote that "living systems ... have numerous
properties that are simply not found in the inanimate world."[8]

> ...I believe that a unification of science is indeed possible if we
> are willing to expand the concept of science to include the basic
> principle and concepts of not only the physical but also the
> biological sciences. Such a new philosophy of science will need
> to adopt a greatly enlarged vocabulary – one that includes such
> words as biopopulation, teleonomy, and progress. It will have to
> abandon its loyalty to a rigid essentialism and determinism in
> favor of a broader recognition of stochastic processes, a plural-
> ism of causes and effect, the hierarchical organization of much
> of nature, the emergence of unanticipated properties at higher
> hierarchical levels, the internal cohesion of complex systems, and
> many other concepts absent from – or at least neglected by – the
> classical philosophy of science.[9]

Mayr emphasized the role of chance in evolution:

> Evolutionary change in every generation is a two-step process:
> the production of genetically new individuals and the selection
> of the progenitors of the next generation. The important role of
> chance at the first step, the production of variability is universally
> acknowledged, but the second step, natural selection, is on the
> whole viewed rather deterministically: Selection is a non-chance
> process. What is usually forgotten is the important role chance
> plays even during the process of selection. In a group of sibs it
> is by no means necessarily only those with the most superior
> genotypes that will reproduce. Predators mostly take weak or
> sick prey individuals, but not exclusively, nor do localized nat-
> ural catastrophes (storms, avalanches, floods) kill only inferior
> individuals. Every founder population [the parent population of

7   Nagel, Thomas, "The Core of 'Mind and Cosmos'," New York Times, August 18, 2013

8   Mayr, E., *Toward a New Philosophy of Biology*. (Cambridge: Harvard University Press, 1988) p. 1.

9   ibid, p. 21.

a new species] is largely a chance aggregate of individuals, and the outcome of genetic revolutions, initiating new evolutionary departures, may depend on chance constellations of genetic factors. There is a large element of chance in every successful colonization. When multiple pathways toward the acquisition of a new adaptive trait are possible, it is often a matter of a momentary constellation of chance factors as to which one will be taken.[10]

Physicalism's picture of the natural order has no room for the kinds of pervasive and objective randomness that evolution requires, according to Mayr and other evolutionary biologists.

Physicalist views of evolution characterize it as a mechanical process – a process whose outcomes are predictable, at least in principle. Some evolutionary biologists argue that evolution is a creative process – one that brings into being novel and unpredictable forms of organized complexity. This was the central theme of the philosopher Henri Bergson's masterwork, *L'Évolution créatrice* (*Creative Evolution*, 1907). Bergson's books combined philosophical and scientific erudition with an engaging and persuasive literary style. They won him a wide and enthusiastic readership. In 1927 he was awarded the Nobel Prize for Literature. But Bergson didn't believe that creative biological processes are consistent with well-established physical laws. He argued that physical laws couldn't govern either evolution or consciousness because they rest on an oversimplified characterization of time. Physics represents time by a line and moments in time by points on the line. In his doctoral dissertation *Time and Free Will* Bergson argued that this static representation deprives time of its essential dynamic character. As time manifests itself in creative biological contexts, including evolution and free human acts, it is a dynamic process, "pure becoming." Evolution, Bergson argued, transcends scientific description. It is driven by an *élan vital* or vital impulse. Because that impulse informs all living organisms, Bergson argued, in *Introduction to Metaphysics*, that we can grasp it through an effort of intuition even if we can't describe it in the language of science.

The *élan vital* didn't survive twentieth-century advances in our understanding of biology. Experimental and observational evidence leaves little room, if any, for doubting that physical laws apply across the board – to living organisms and physical systems alike. Biology's

10   Mayr, E. 1983. *The American Naturalist* 121: 324-33; reprinted in *Toward a New Philosophy of Biology*. pp. 148-159.

current picture of the living world also dispenses with "biotonic" laws – laws that apply only to living organisms.

But while Bergson's interpretation of evolution is no longer tenable, his description of it as a creative process remains compelling. Embracing that description while rejecting Bergson's metaphysics, Theodosius Dobzhansky, like Mayr a principal architect of current evolutionary theory, argued that evolution is indeed "a creative process, in exactly the same sense in which composing a poem or a symphony, carving a statue, or painting a picture are creative acts":

> Can the word 'creative' be validly applied to a process that has no foresight and no ability to devise means to a chosen goal? ...Evolution [like artistic creation] not only brings novelties into being, but these novelties present embodximents of new ways of life. ...Every new form of life that appears in evolution can, with only moderate semantic license, be regarded as an artistic embodiment of a new conception of living.[11]

Like artistic creativity, evolutionary creativity requires randomness, or indeterminism, to be an objective feature of the natural world. Quantum indeterminism isn't enough. If Mayr's and Dobzhansky's views of evolution are correct, macroscopic indeterminism – the prevalence of chance at macroscopic levels of description – must be a pervasive and objective feature of the macroscopic world, as it seems to be of the world of experience. But it has no place in physicalism's deterministic picture of physical reality, as described by Pierre Simon Laplace in 1814:

> We may regard the present state of the universe as the effect of its past and the cause of its future. An intellect which at a certain moment would know all forces that set nature in motion, and all positions of all items of which nature is composed, if this intellect were also vast enough to submit these data to analysis, it would embrace in a single formula the movements of the greatest bodies of the universe and those of the tiniest atom; for such an intellect nothing would be uncertain and the future just like the past would be present before its eyes.[12]

11   Dobzhansky, Theodosius.. "Chance and Creativity in Evolution" in *Studies in the Philosophy of Biology*, edited by F.J. Ayala and T. Dobzhansky. (Berkeley: University of California Press 1974) p. 329.

12   Laplace, Pierre Simon, *A Philosophical Essay on Probabilities*, translated into English from the original French 6th ed. by Truscott, F.W. and Emory, F.L., (New York, Dover Publications 1951) p.4.

In his essay "Chance" (1908) the mathematician, physicist, and philosopher Henri Poincaré reaffirmed this worldview:

> Every phenomenon, however trifling it be, has a cause, and a mind infinitely powerful and informed concerning the laws of nature could have foreseen it from the beginning of the ages. If a being with such a mind existed, we could play no game of chance with him; we should always lose.

Yet in Mayr's and Dobzhansky's views of evolution, all living organisms are winners in games of chance whose outcomes not even "a mind infinitely powerful and informed concerning the laws of nature" could have foreseen." Physicalism equates chance to imperfect human knowledge of laws or initial conditions. By contrast, Mayr's and Dobzhansky's views of evolution require chance events to be unpredictable in principle. Physical laws allow us to predict future states of complex systems if we know enough about their present state. By contrast, evolution gives rise to forms of complex order that aren't prefigured or implicit in earlier states of the universe.

Randomness is synonymous with disorder. Its antithesis, in biological contexts, is organized complexity. Because physicalism's account of physical reality doesn't accommodate macroscopic randomness, neither does it accommodate organized complexity and the processes that give rise to it (creative processes). A scientific worldview that accommodated consciousness and evolution, as conceived by Mayr, Dobzhansky, and other contemporary evolutionary biologists, would also allow us to characterize randomness and organized complexity in an expanded language of physics.

This book describes such a worldview and its underpinnings.

II
Cosmological Assumptions

Natural science's account of the world combines a small set of strongly confirmed physical laws with supplementary conditions that characterize more or less idealized models of physical systems to which the laws apply. Supplementary conditions are of two kinds: *initial conditions* characterize a system at a particular moment; *boundary conditions* describe the system's interaction or absence of interaction with its environment. Each of our strongly confirmed laws applies to many possible systems and processes. For example, astronomers apply Newton's law of gravitation and his laws of motion to idealized models of astronomical systems ranging from the Earth-Moon system to clusters of galaxies. A handful of deep, abstract, and strongly confirmed physical laws govern an unlimited variety of models of system and processes characterized by freely chosen initial and boundary conditions. A theory's testable predictions serve to test both its laws and its initial conditions.

The most fundamental supplementary conditions are cosmological assumptions – assumptions about both the physical universe and the laws. Physicalism assumes that our strongly confirmed physical laws are correct – or at least excellent approximations to still deeper and more highly unified laws. But physicalism isn't a consequence of this assumption alone. It depends also on the assumption that at each moment the physical universe is in a definite physical state:

Physicalism's underlying assumption. A complete description of the physical universe assigns definite values to the physical quantities that characterize macroscopic systems. Thus it fully describes and individuates every macroscopic physical system.

Together with the fact that our strongly confirmed physical laws link a description of any given state of an undisturbed system to which the laws apply to descriptions of the system's earlier and later states, this assumption, implies that the outcomes of nearly all physical processes – indeed all processes other than quantum measurements – are predictable in principle. In other words, the assumption allows us to pass from the (uncontested) proposition that our strongly confirmed physical laws are deterministic – that they link the present state of any undisturbed system to which the laws apply to any of the system's past or future states – to the proposition that these laws plus initial and boundary conditions that conform to physicalism's underlying assumption determine future states of undisturbed systems, including the universe itself. It allows us to pass from the determinism of physical laws to cosmic determinism.

Of course, experiments and observations don't provide the exact values of physical quantities that have continuous ranges of possible values, like position, mass, and temperature. Random measurement errors – a consequence of human ignorance – locate the values of such quantities within small subranges. Improved experiments and observations reduce these subranges. And since measurement errors can't be eliminated entirely, the measured values of physical quantities can't distinguish the Sun from other suns or the Galaxy from other galaxies. Which is why both Laplace's and Poincaré's definitions of determinism, quoted above, refer to an imaginary being that knows the precise values of all relevant physical quantities.

The universe envisaged by Newton, Laplace, and Poincaré is made up of particles of various sizes and shapes, moving and interacting in ways governed by Newton's laws of motion. In such a universe every particle (or its center of mass) has a definite position at each moment, defined by a set of three real numbers – the particle's rectangular coordinates in a fixed but arbitrary coordinate system. But cosmologists now agree that the early universe wasn't a sea of classical particles. It was a sea of particles governed by quantum laws. One strongly confirmed prediction of these laws is that two particles of the same kind, such as a pair of electrons or neutrinos or photons, are indistinguishable in a way that has no classical counterpart. Two classical particles may have identical intrinsic properties, such as mass, electric charge, and spin. Yet they are nevertheless distinguishable: one is here, the other is there. By contrast, the laws of quantum physics don't assign each member of an assembly of elementary particles of the same kind its own position or, more generally, its own single-particle state. The assembly has distinct quantum states, but the individual particles that compose it do not. As a result, states of the early universe *could* be fully described in ways that don't privilege any spatial position or direction. For example, a state of the early universe *could* be a state fully characterized by its temperature and the relative concentrations of elementary particles.

If the initial conditions that characterize the early universe don't privilege any point or direction in space, the same will be true of the conditions that characterize the universe at later times, because the laws that link earlier to later states don't introduce a privileged position or direction. So a complete description of the physical universe can't contain descriptions of this star or that galaxy. Instead it describes what I'll call *cosmological ensembles* – infinite collections of near-replicas uniformly and isotropically distributed in space. Thus geophysics isn't just about Earth; it is about an infinitely dispersed collection of planets whose observable properties are indistinguishable from those of Earth.

I'll refer to the assumption that a complete description of the universe doesn't privilege any spatial position or direction as *the strong cosmological principle*. It is a strengthened version of an assumption that underlies most contemporary theories (and observations) of the astronomical universe:

> *The cosmological principle.* There are coordinate systems relative to which a statistical description of the distribution of matter and motion in the universe at any given moment doesn't privilege any spatial position or direction.

The strong version of this assumption replaces the word *statistical* by the word *complete*:

> *The strong cosmological principle.* There are coordinate systems relative to which a *complete* description of the distribution of matter and motion in the universe at any given moment doesn't privilege any spatial position or direction.

The strong cosmological principle implies that, contrary to the assumption underlying physicalism, a complete description of the physical universe doesn't describe individual physical systems such as the Sun and Earth. Instead it describes infinite collections of replicas – *cosmological ensembles*. The systems that make up an ensemble are uniformly and isotropically distributed in space. The ensemble itself is fully characterized by a set of probabilities, or *probability distribution*, of its members' physical attributes. The probability that an attribute has a given value or a value that lies in a given range of values has an objective interpretation: it is the fraction of replicas in a cosmological ensemble in which the attribute has that value or a value in the given range.

The strong cosmological principle relies on quantum physics. It couldn't hold in a universe made up of particles governed by the laws of classical physics. Consider the simplest model of such a universe: a random, statistically uniform distribution of identical particles. In such a universe the distance between any given particle and (say) its nearest neighbor, measured in a fixed unit of length, is a real number – a number represented by a point on the number line. This number characterizes the particle uniquely, because the probability that a second particle has the same distance from its nearest neighbor is zero. (As the mathematician Georg Cantor showed in 1891, there are infinitely more real numbers in any real-number interval than there are particles in the universe.) The picture of the natural world described in this book rests on a pair of related cosmological assumptions. The first is the strong cosmological principle. The second is the *assumption of primordial randomness*:

> *Primordial randomness.* The probability distributions that characterize the earliest state of the astronomical universe

to which our present physical laws apply are maximally random.

What does *randomness* mean in this context? As discussed in more detail below, every probability distribution has two complementary attributes: *randomness* and *information*.

Randomness is closely related to – but not identical with – *entropy*. Between 1850 and 1865 Rudolf Clausius discovered a previously unnoticed property of effectively isolated macroscopic systems that have relaxed into a macroscopically uniform, unchanging state called *thermal equilibrium*. He named this property *entropy* and proved that the second law of thermodynamics (the science of heat and its transformations) is equivalent to the statement that the entropy of a closed, or isolated, system never decreases.

Although randomness is a property of any probability distribution, Ludwig Boltzmann introduced it, in 1872, in a particular physical context. He showed that it is the counterpart of Clausius's entropy in the molecular theory of gases – a theory that seeks to found thermodynamics on what was then the speculative assumption that a gas is a collection of particles whose motions and interactions are governed by Newton's laws of motion. Boltzmann's H theorem states that the randomness of an isolated sample of an ideal gas never decreases.

In wider contexts randomness is a measure of disorder. Its complementary property, *information*, is the amount by which a probability distribution's information falls short of its largest possible value. It is a measure of order.

The assumption of primordial randomness contradicts a proposition advanced by Rudolf Clausius in 1865: "The entropy of the world tends toward a maximum." This statement has been widely accepted. As astrophysicist Arthur S. Eddington explained in an influential book, *The Nature of the Physical World*,

> The practical measure of the random element which can increase in the universe but never decrease is called entropy. ...Entropy continually increases. ...The law that entropy always increases – the second law of thermodynamics – holds, I think, the supreme position among the laws of Nature. If someone points out to you that your pet theory of the universe is in disagreement with Maxwell's equations – then so much for Maxwell's equations. If it is found to be contradicted by observation – well, these experimentalists do

bungle things sometimes. But if your theory is found to be against the second law of thermodynamics I can give you no hope; there is nothing for it but to collapse in the deepest humiliation.[1]

More recently the mathematician Roger Penrose likewise endorsed Clausius' formulation of the law, arguing that the universe began to expand from a highly organized state of exceedingly low entropy and has been becoming progressively more disorganized.[2] But let's take a closer look. The first of thermodynamics' two main laws says that heat and mechanical energy, or work, are interconvertible at a fixed rate of exchange. The second law imposes restrictions on the conversion of heat into work. In one of its forms it states that no cyclic heat engine can convert heat drawn from a heat reservoir at a fixed temperature entirely into work. Clausius defined the entropy of a macroscopic system in terms of its changes when a macroscopic system accepts heat from or delivers heat to an external heat reservoir. He showed that the Second Law is equivalent to the statement that the entropy of a closed (or undisturbed or isolated) system never decreases. The two laws underpin predictions about all macroscopic processes involving heat flow, including the life-sustaining energy transactions between living organisms and their environments.

Both have been strongly confirmed.

At first sight it may seem easy to generalize the definition of entropy and the law of entropy change from macroscopic systems to the universe. Seemingly, we can represent the universe as a collection of weakly interacting macroscopic systems for each of which entropy is a well-defined quantity that never decreases. To calculate the entropy of the universe we simply add up these contributions. The resulting quantity, the "entropy of the universe", never decreases. But, as Enrico Fermi emphasized in his *Lectures on Thermodynamics* and as we'll discuss in more detail below, this recipe works "only if the energy of the system is the sum of the energies of all the parts and if the work performed by the system during a transformation is equal to the sum of the amounts of work performed by all the parts."[3] The systems experimental physical scientists and biologists study usually

---

1   Eddington, Arthur S. *The Nature of the Physical World*, (Ann Arbor: University of Michigan Press 1958) p. 74

2   Penrose, Roger. *The Emperor's New Mind*. (Oxford: Oxford University Press 1989) Chapters 7, 8

3   Fermi, Enrico. *Thermodynamics*, (New York: Prentice-Hall 1937) p. 53

satisfy these conditions. But astronomical systems don't satisfy the first condition because the gravitational potential energy of a system held together by the mutual gravitational attraction of its parts isn't equal to the sum of the parts' gravitational potential energies. For example, the gravitational potential energy of a uniform gas sphere of given mass density is proportional not to its mass but to the square of its mass.

That Clausius's generalization of his law doesn't apply to self-gravitating systems shows up in the fact that self-gravitating systems have negative heat capacity: the mean temperature of a self-gravitating gas cloud increases when heat, in the form of radiation, *leaves* the cloud. The law of entropy growth implies, however, that any system to which the law applies has positive heat capacity: its mean temperature increases when heat flows into it.

Since the universe is made up of self-gravitating systems, it doesn't satisfy the additivity condition. So we can't extend the law of entropy non-decrease to the universe as a whole.

By contrast, any probability distribution that characterizes the state of the universe has a well-defined randomness. So it does make sense to postulate that the randomness of a statistical description of the universe tends toward a maximum. I think this assumption is implicit in the physicalist picture of the natural world, which contains only the entities mentioned in physical theories, and I think that may be how Eddington and Penrose interpreted Clausius's dictum about the entropy of the universe.

This book argues that a picture of the natural world based (in part) on the alternative assumption that the universe has expanded from a primordial state of zero information or maximum randomness is more inclusive than physicalism's world picture. I'll argue that it accommodates not only consciousness but also creative processes, including free will and biological evolution itself.

The case for the two cosmological assumptions rests in part on the claim that they characterize the physical universe as simply as possible. I argue below that these assumptions throw light on three longstanding problems in physics: the problem of time's arrow – the apparent conflict between the irreversibility of macroscopic processes involving heat flow and the fact that the underlying microscopic laws do not distinguish between the direction of the past and the direction of the future; the measurement problem of quantum mechanics – the apparent conflict between the indeterminacy of quantum measurement outcomes and the determinacy of the physical laws that govern the physical processes involved in measurements; and the interpretation of probabilities in the statistical theories that link macrophysics and microphysics.

III

The Universe's Large-Scale Structure

Isaac Newton envisioned the universe as an infinitely extended assembly of particles, each of which is in a definite place at each moment. Replying to a question from the theologian Nicholas Bentley, who was preparing a set of lectures intended to defend religion from atheism, Newton, a devout theist, wrote, in 1692:

As to your first query, it seems to me that if the matter of our sun and planets and all the matter in the universe were evenly scattered throughout all the heavens, and every particle had an innate gravity toward all the rest, and the whole space throughout which this matter was scattered was but finite, the matter on the outside of the space would, by its gravity, tend toward all the matter on the inside, and by consequence, fall down into the middle of the whole space and there compose one great spherical mass. But if the matter was evenly disposed throughout an infinite space, it could never convene into one mass; but some of it would convene into one mass and some into another, so as to make an infinite number of great masses, scattered at great distances from one to another throughout all that infinite space. And thus might the sun and fixed stars be formed, supposing the matter were of a lucid nature. But how the matter should divide itself into two sorts, and that part of it which is fit to compose a shining body should fall down into one mass and make a sun and the rest which is fit to compose an opaque body should coalesce, not into one great body, like the shining matter, but into many little ones; or if the sun at first were an opaque body like the planets, or the planets lucid bodies like the sun, how he alone would be changed into a shining body whilst all they continue opaque, or all they be changed into opaque ones whilst he remains unchanged, I do not think explicable by mere natural causes, but am forced to ascribe it to the counsel and contrivance of a voluntary Agent.

Newton doesn't mention this cosmogonic hypothesis in his masterwork, *The Mathematical Principles of Natural Philosophy* (the *Principia*), which had appeared five years earlier. In *Relativity, the Special and the General Theory,* a popular exposition of relativity, Einstein points out that Newton's model of the astronomical universe, in which matter is distributed more or less uniformly throughout infinite Euclidean space, is inconsistent with his (Newton's) law of gravitation:

> [Newton's theory] requires that the universe should have a kind of center in which the density of stars is a maximum, and that as we proceed outwards from this center the group density of the stars should diminish, until finally, at great distances, it is succeeded by an infinite region of emptiness. The stellar universe ought to be a finite island in the infinite ocean of space.[1]

Einstein appends a proof of this claim:

The proof rests on the fact that Newton's law of gravitation can be expressed as a picture of "lines of gravitational force." Every particle is the terminus of such lines. Their number is a fixed multiple of the particle's mass. If mass were uniformly distributed in infinite Euclidean space, the number of lines of force passing through unit area of any given sphere – and hence the magnitude of the gravitational force at any point on the sphere – would be proportional to the sphere's radius and would therefore increase without limit as the radius increased, "which [Einstein writes] is impossible."

Einstein hoped that his own theory of spacetime structure and gravitation *would* apply to Newton's simplified model of the physical universe – an infinitely extended, uniform distribution of mass/ energy. In 1917, two years after his account of the general theory of relativity appeared in print, he summarized his efforts to construct a description of the universe based on that model. He concluded that the equations that link the structure of spacetime to its contents – the theory's "field equations" – do *not* have a solution that describes a motionless, infinitely extended, uniform distribution of mass/energy.

Einstein searched for a way to reconcile general relativity with his preferred model of the astronomical universe. General relativity

1   Einstein, A., *Relativity, the Special and the General Theory,* 15th edition, (Crown, New York, 1952) p. 106.

incorporates Bernhard Riemann's mathematical theory of curved n-dimensional continuums. That theory allows a three-dimensional space to be curved both globally and locally, like a model of Earth's surface. If physical space had constant positive curvature it would be the three-dimensional "surface" of a four-dimensional ball. Like a sphere (the two-dimensional surface of a three-dimensional ball), such a space is unbounded – it has no edge – yet finite. Disappointingly, the field equations seemed to have no solution that describes a finite quantity of matter spread evenly throughout a finite space of constant positive curvature. Convinced that a plausible model of the universe must be uniform and isotropic apart from local irregularities, Einstein now took a step he later regretted. He modified the field equations of his 1915 theory in a way that would allow them to have such solutions. The 1915 theory links the structure of space-time to the distribution of mass/energy in a way that involves only a single physical constant – the constant that appears in Newton's law of gravitation. The field equations of the modified theory contain an additional term, whose coefficient is a second constant, the *cosmological constant*.

Five years later, in 1922, the mathematician Alexander Friedmann wrote a paper claiming that Einstein's original field equations do in fact have solutions that describe uniform, unbounded distributions of mass – solutions Einstein had overlooked. Einstein's mistake, Friedmann claimed, had been to assume that such a distribution must be *static*. True, there was as yet no observational evidence that the distribution of matter was not static on large scales, though there soon would be. But Friedmann was a mathematician, not a physicist or astronomer. So he asked himself: Do Einstein's original field equations have solutions that describe a uniform, unbounded, but *non-static* distribution of mass? He found that they do. The mass distribution may be infinitely extended or, if space has positive curvature, it may occupy a finite volume. If the medium is infinitely extended, a solution of the field equations describes a universe that expands forever from a state of nominally infinite density in the finite past. If the distribution of mass/energy and the space it occupies are finite, a solution (of the same field equations) cycles endlessly through alternate periods of expansion and contraction. These solutions of Einstein's 1915 field equations describe models of the physical universe that, so far as I know, no astronomer, philosopher, or writer of science fiction had previously described. At first Einstein rejected Friedmann's reasoning, but on reflection he admitted he had been mistaken. He wholeheartedly embraced Friedmann's solution of the cosmological problem and urged the editors of the journal to which Friedmann had submitted his paper to accept it (which, of course, they did). Einstein also decided that it had been a mistake to add an extra term to his original field equations. Ironically, most contemporary cosmologists believe the extra term is needed to account for astronomical evidence.

"The Realm of the Nebulae"

Using a telescope he himself had designed and built, Galileo discovered in 1610 that the Milky Way, a diffuse band of light that encloses a great circle on the dome of the sky, is made up of myriad "suns" too faint for the naked eye to resolve.

In 1750 Thomas Wright, an instrument maker, suggested that just as the Earth's orbit lies near the central plane of a flattened system of planets circling the Sun, the stars that make up the Milky Way belong to a flattened system of Sun-like objects whose members, including the Sun, circle a distant center.

Immanuel Kant embraced Wright's hypothesis and, in his 1755 book *Universal Natural History and Theory of the Heaven*, greatly extended it. Kant suggested that the stars are suns, many of them surrounded by flattened systems of planets, which, like Jupiter and Saturn, are themselves surrounded by flattened systems of satellites; that not only the Milky Way but also the "nebulae" – elliptical cloud-like objects scattered among the stars – are very distant stellar systems with the same structure as planetary systems – a massive central object surrounded by a disc-like family of satellites, all revolving about the central object in the same sense; and that these stellar systems themselves belong to still larger unbounded system with the same structure.

Kant argued that all these systems could have formed by a single process: a cloud of dispersed matter collapses under the combined action of Newton's gravitational attraction and a hypothetical repulsive force. (Curiously, because he was an admirer and expounder of Newtonian physics, Kant didn't recognize that Newton's theory makes a repulsive force superfluous. As the mathematician Pierre

Simon Laplace (1749 – 1827) later showed, unless a cloud has zero angular momentum, or spin, it collapses into a flattened system in which the centrifugal, or radially outward, acceleration of a particle circling the central mass balances the radially inward gravitational acceleration of the particle produced by matter inside the particle's orbit.) Astronomers disagreed about whether the "nebulae" are relatively nearby gas clouds or distant stellar systems. In 1920 astronomers Harlow Shapley and Heber Curtis held a famous debate on this question. Curtis defended Kant's view that the nebulae are stellar systems comparable in all respects to our own system, the Milky Way. Shapley argued that they are gas clouds in the Milky Way.

In the 1920s the astronomer Edwin Hubble and his colleagues, using what was then the world's most powerful telescope, the recently completed 100-inch reflecting telescope at the Mount Wilson Observatory in southern California, made systematic observations of the nebulae that resolved the debate. Longexposure photographs resolved bright stars in the largest and brightest of the nebulae, the Andromeda Nebula, and analyses of photographic and spectroscopic data confirmed that it was a stellar system comparable in size and stellar composition with our own stellar system, thus confirming Kant's hypothesis. The nebulae were indeed galaxies (a term suggested by Shapley). In *The Realm of the Nebulae* Hubble wrote:

> Investigations of the observable region as a whole have led to two results of major importance. One is the homogeneity of the region – the uniformity of the large-scale distribution of the nebulae. The other is the velocity-distance relation. [2]

The first result depended on a discovery by the astronomer Henrietta Swan Leavitt (1868-1921). Some stars are close enough to have measurably different directions when the Earth is on opposite sides of the Sun. From measurements of this difference, the star's parallax, one can deduce its distance, much as Thales deduced the distance of a ship at sea from measurements of its parallax and the measured distance between the two points of observation. Astronomers then deduce the star's true brightness, or luminosity, from its apparent brightness and the fact that an object's apparent brightness diminishes like the reciprocal of the square of its distance. Conversely, if an astronomer knew a celestial object's luminosity she could infer

---

2   Hubble, Edwin, *The Realm of the Nebulae*, (Yale University Press, 1936).

its distance from a measurement of its (apparent) brightness. Leavitt studied 1777 variable stars in the Magellanic Clouds, a pair of dwarf galaxies that orbit our own galaxy. Among these variable stars were Cepheids, pulsating stars whose brightness varies periodically. She found that the mean brightnesses of the Cepheids she studied were tightly correlated with their periods. Since stars in the Magellanic Clouds are all at nearly the same distance from Earth, Leavitt's discovery meant that the luminosities of the Cepheids she studied were tightly correlated with their periods.

To calibrate Leavitt's period-luminosity relation astronomers use Cepheids that are close enough to have measurable parallaxes. They then use the calibrated relation between distance and luminosity to infer the distances of more distant Cepheids from their (apparent) brightnesses. In 1923, two years after Leavitt's death, Hubble discovered the first of several Cepheids in the Andromeda Nebula, the brightest of the external galaxies. Its measured brightness and Leavitt's period-luminosity relation played the decisive role in establishing that the Andromeda Nebula is a stellar system comparable in size and stellar composition with our own galaxy.

Hubble's assumption that Cepheid-like variables in external galaxies satisfy the same relation between true brightnesses and period of variation as Cepheids in our own galaxy exemplifies an assumption he called "the principle of uniformity." Using that assumption to interpret measurements of the brightnesses of galaxy images on photographs made with the 100-inch reflector, Hubble concluded:

> There is no evidence of a thinning-out, no trace of a physical boundary. There is not the slightest suggestion of a supersystem of nebulae isolated in a larger world. Thus, for the purposes of speculation, we may apply the principle of uniformity, and suppose that any other equal portion of the universe, selected at random, is much the same as the observable region. We may assume that the realm of the nebulae is the universe and that **the observable universe is a fair sample.** (emphasis added)

In Hubble's hands the assumption that "any other equal portion of the universe, selected at random, is much the same as the observable region" proved to be an invaluable research tool. As he understood, "much the same" doesn't mean "exactly the same." Hubble's successors have devoted much time, effort, and ingenuity to refining and extending the principle of uniformity. For example, they have discovered that there are different kinds of Cepheids, with slightly but significantly different period-luminosity relations, and that some of the differences depend on systematic differences in chemical composition between different stellar populations. Such differences, however, are consistent with a more general cosmological assumption, mentioned earlier: the *cosmological principle*:

> Relative to a suitably defined system of spacetime coordinates, the average properties of galaxies and their distribution in space at any given moment are the same everywhere and in all directions.

Or more briefly:

> *The astronomical universe is statistically homogeneous and isotropic.*

The second of Hubble's "two results of major importance" was the velocity-distance relation. Whereas the first result (that the "external nebulae", or galaxies, are stellar systems comparable to our own) depended on brightness measurements, the second depended on measurements of the displacements of absorption lines in the spectra of distant galaxies.

Astronomical Spectroscopy

By passing a beam of sunlight through a glass prism, Newton separated it into what he called "colored rays." He found that differently colored rays were deflected through different angles when they passed through the prism. Newton's successors established that the colored rays are electromagnetic waves with definite frequencies, ranging from 430×10^{12} cycles per second at the red end of the spectrum to 770×10^{12} cycles per second at the violet end.

The spectrum of sunlight is crossed by thousands of dark lines, each of which results from the absorption and subsequent isotropic reemission of light by a specific kind of atom or molecule. The lines produced by a single atom or molecule are very narrow, but those produced by a macroscopic gas sample – a portion of the solar atmosphere, for example – are broadened in wavelength by the Doppler effect associated with the relative line-of-sight motions of the atoms or molecules in the sample. (The Doppler effect increases the frequency of monochromatic light emitted by an approaching source and decreases the frequency of light emitted by a receding source. The fractional change in frequency is proportional to the velocity of approach or recession.)

The spectrum of a distant galaxy superimposes contributions from the galaxy's stars (and other emitters of light). The breadth of an absorption line measures the dispersion of the relative line-of-sight velocities of the stars. The displacement of the line's center from its rest-value measures the line-of-sight velocity of the galaxy's center of mass (or, more accurately, "center of light"), relative to the telescope.

Extending pioneering measurements of galaxy redshifts by Vesto Slipher, Hubble and his colleague Milton Humason found that the absorption lines in the spectra of very distant galaxies were always displaced toward the red end of the spectrum. Moreover, this displacement, or red shift, had a systematic component, proportional to the galaxy's distance: the points on a graph of measured redshift against estimated distance hugged a straight line. And because the amount of scatter didn't increase systematically with estimated distance, Hubble could attribute deviations from the straight-line relation to gravitational accelerations arising from local non-uniformities in the spatial distribution of galaxies – nonuniformities whose average properties didn't depend on distance from the observer.

The cosmological principle implies that there's nothing special about our galaxy's place in the universe of galaxies. It also implies that the population of galaxies extends indefinitely in all direction. The proportionality between the average line-of-sight velocity of a galaxy and its distance from the observer then has a straightforward interpretation:

> The unbounded distribution of galaxies is undergoing a
> uniform expansion – an expansion that looks the same
> from every vantage point.

Hubble announced the velocity-distance relation in 1929. At the time, he didn't know that his discovery confirmed Alexander Friedmann's seven-year-old prediction, based on the general theory of relativity and the assumption that matter is more or less evenly distributed in an unbounded universe.

The Early Universe

In an expanding universe the mass density, averaged over scales greater than the largest scales on which the mass distribution is non-uniform, decreases steadily with time. So as we look back in time, we see the average mass density steadily increasing. Eventually it becomes equal to, and then surpasses, the average mass density of galaxy clusters. At these and earlier times galaxy clusters couldn't yet have existed. As we look back still further in time, we reach epochs in which galaxies themselves couldn't have existed. Still earlier, the average cosmic mass density exceeded stellar densities. Even earlier, the average mass density would have been so high that atoms and molecules couldn't have existed. Finally, at the earliest times at which our current physical theories apply, the universe would have been a more or less uniform distribution of elementary particles and photons. This conclusion seems inescapable if we assume that the average properties of the astronomical universe are the same everywhere and that quantum theory and Einstein's theory of gravitation apply in their respective domains.

In 1964 physicists Arno Penzias and Robert Wilson at the Crawford Hill location of Bell Telephone Laboratories discovered what came to be called the cosmic microwave background. With a radio telescope they had built to observe electromagnetic radiation emitted by astronomical objects with wavelengths ranging from tenths of a centimeter to a few centimeters, they detected a diffuse radiation field whose intensity never varied and was the same in all directions. The spectrum of the radiation was indistinguishable from that of radiation in a cavity whose walls are maintained at a temperature around three degrees above absolute zero (on the Kelvin scale). Later more accurate measurements confirmed this conclusion and placed the temperature at 2.725 K.

In the most widely accepted model for the origin of the cosmic microwave background, the temperature of the background radiation has steadily decreased from very high values near the beginning of the cosmic expansion:

> At the time when [the temperature] T $\approx 10^{12}$, the universe contained photons, muons, antimuons, electrons, positrons, neutrinos and antineutrinos. In addition, there was a very small nucleonic

contamination, with neutrons and protons in equal numbers. All of these particles were in thermal equilibrium.[3]

In other models the primordial universe was cold, and the cosmic microwave background came into being later.[4]

Quantum Indistinguishability

Elementary particles and photons come under the jurisdiction of quantum mechanics. A statistically uniform distribution of elementary particles and photons in thermal equilibrium differs profoundly from a statistically uniform and random distribution of particles governed by the laws of classical, or non-quantum, physics. A statistically uniform and random distribution of classical particles with given mass density and temperature has infinitely many microscopically distinct realizations, or microstates. But a statistically uniform distribution of identical particles governed by quantum mechanics is fully characterized by the distribution's mass density and its temperature. It has a single microstate. This conclusion follows from a deep, consequential, and non-intuitive feature of the quantum world: *quantum indistinguishability*. I'll argue shortly that quantum indistinguishability underpins this essay's main argument – *that randomness is an objective and pervasive feature of the physical universe*.

What is quantum indistinguishability? Two classical particles may have identical properties, such as mass, electric charge, and spin. Yet they are nevertheless distinguishable: one is here, the other is there. Now consider a statistically uniform assembly of identical classical particles. At any given moment the distance between a given particle (or its center of mass) and the particle's nearest neighbor, expressed as a multiple of a fixed unit of distance, is a real number – a number represented by a point on the number line or by a non-terminating decimal. The probability that any other particle in the assembly has the same distance from its nearest neighbor is zero. For the probability that the first n digits in the decimal expansions of two randomly selected real numbers coincide is $1/10^n$, which approaches zero as n increases without limit. So each member of an infinitely extended, statistically uniform and random distribution of identical particles is distinguishable in principle from all other members by the relative positions of its neighboring particles (or even of one of

---

3   Weinberg, S. 1972. *Gravitation and Cosmology*. New York: Wiley, p. 528.
4   Aguirre, A. 1999. Astrophysical Journal, 521, pp. 17-29.

them). Such an assembly is in a definite microstate at each moment; and the number of possible microstates is infinite in the same way that the number of real numbers in an interval of the number line is infinite.

Like classical particles of the same kind, quantum particles of the same kind have identical intrinsic properties – mass, electric charge, spin, and magnetic moment. But they are also indistinguishable in a more radical way that depends on a feature of the mathematical description of quantum states that has no counterpart in experience or in pre-quantum physics. This new kind of indistinguishability – quantum indistinguishability – strongly influences observable, macroscopic properties of matter and light. For example, it is behind the fact that atoms in the same column of the periodic table of chemical elements (such as hydrogen, lithium, sodium, and potassium) have similar chemical properties (for example, the named elements are all monovalent). It is also behind the fact that the distribution of photon energies in a box whose walls are maintained at a fixed temperature differs from the distribution of the kinetic energies of gas particles in the same box.

To understand quantum indistinguishability you need to know two things about the grammar of state vectors, the mathematical objects that represent quantum states. (1) Two state vectors that differ only in sign represent the same quantum state. (2) Exchanging the labels of two particles in a state vector that represents the state of an assembly of identical particles must either leave the state vector unaltered or change its sign (that is, multiply it by –1). These are the only possibilities the rules of quantum mechanics allow. They define two classes of elementary particles, called *fermions* (after Enrico Fermi) and *bosons* (after Satyendra Nath Bose). Electrons, protons, and neutrinos are fermions – the joint state vector of an assembly of fermions changes sign when you exchange the labels of two particles in the assembly. Helium atoms in their states of lowest energy and photons are bosons. The joint state vector of an assembly of bosons doesn't change when you exchange the labels of two particles in the assembly. Which of the two classes an elementary particle belongs to depends on its spin. Particles of spin 1/2, 3/2, ... in units of $h/2\pi$ are fermions, particles of spin 0, 1, 2, ... in the same unit are bosons.

An assembly of fermions has different observable properties from an assembly of bosons, and both collections have different observable properties from an assembly of identical classical particles. Some examples:

— From plausible but then-controversial statistical assumptions James Clerk Maxwell, in 1860, deduced the distribution of molecular kinetic energies in a gas sample in thermal equilibrium. He found that each molecular-velocity component has a "normal distribution," represented by the bell-shaped curve that represents the distribution of random measurement errors, and that the width of the distribution is proportional to the temperature of the gas sample. Much later, experiments confirmed his prediction.

— In 1900 Max Planck devised a formula that closely fitted recently improved measurements of the spectrum (frequency distribution) of thermal radiation (light in a box whose walls are maintained at a fixed temperature). In view of Einstein's photon hypothesis (now a firmly entrenched feature of quantum physics) we can think of Planck's law as describing the equilibrium distribution of photon energies. It differs markedly from Maxwell's law for the distribution of the energies of classical particles in thermal equilibrium.

— Two fermions can't occupy the same single-particle state. Wolfgang Pauli discovered an instance of this rule in 1925. To explain the periodic structure of the periodic table of chemical elements he made two proposals. He suggested that four, rather than three, quantum numbers characterize the state of an electron in an atom. (The first three quantum numbers correspond to the fact that we need three numbers to specify a classical particle's orbit. The fourth quantum number, which had two possible values, turned out to characterize the component of the electron's spin along a fixed direction.) Pauli's second proposal was the rule, known as Pauli's exclusion principle: two electrons can't occupy the same quantum state.

— Because several bosons can occupy the same single-particle states, the members of a collection of bosons in thermal equilibrium (a dilute gas of helium atoms, for example) crowd into the single-particle state of lowest energy at temperatures close to absolute zero.

The Strong Cosmological Principle

The cosmological principle is an assumption about the large-scale structure of the physical universe. It says that there are spacetime coordinate systems in which a description of the universe phrased entirely in terms of probability distributions and average values doesn't discriminate between positions in space or between directions in space. For example, at each moment the mass density, averaged over a sufficiently large region, has the same value everywhere. So too does the average value of the squared difference between the mass density at a point and its (position-independent) average value.

All coordinate systems in which a statistical description of the universe doesn't discriminate between spatial positions or directions have a common time coordinate. In this respect the cosmological principle may seem to reinstate Newton's universal time. In Newton's physics time, and with it the notion of rest, are "absolute" because his laws of motion hold in, and only in, particular spacetime coordinate systems all of which are at rest relative to one another. Einstein's special theory of relativity (1905) abolished absolute time and absolute rest. It demanded that physical laws in no way discriminate between a spacetime coordinate system in which Newton's laws hold and any other coordinate system whose motion is unvarying in speed and direction relative to the first system. Coordinate systems in which Newton's laws hold are called inertial systems. The principle of special relativity requires physical laws to take the same form in all inertial coordinate systems if they are written in a mathematical language devised by Einstein's former teacher, Hermann Minkowski. Minkowski spacetime replaces the Euclidean space plus absolute time of Newton's theory. Newtonian physics becomes a limiting case of special-relativity physics, approximately valid for particle speeds much smaller than the speed of light in empty space. Beginning in

1928 with P.A.M. Dirac's relativistic generalization of Schrödinger's equation, discussed in more detail below, special relativity became a pillar of quantum theory.

The general principle of relativity revokes the privileged status of inertial spacetime coordinate systems. It requires physical laws to take the same form in all coordinate systems that assign the same squared spacetime interval (the squared time interval minus the squared distance interval) to every pair of neighboring point events. As Einstein explains in his popular exposition Relativity, the Special and the General Theory,[5] in his more technical lectures The Meaning of Relativity,[6] and in his comprehensive journal article The Foundations of the General Theory of Relativity,[7] a consistent working-out of the general principle of relativity leads to a unique theory of gravitation and spacetime structure. (The path wasn't easy, though, even for Einstein. One wrong turn, known as the "hole problem" put him off course for two years.[8])

The existence of a preferred family of spacetime coordinate systems for the universe as a whole doesn't clash with the general principle of relativity. The latter constrains the *laws* governing spacetime structure and gravitation; the cosmological principle characterizes a particular system to which the laws apply: the astronomical universe.

The assumption that there exists a system of spacetime coordinates relative to which no statistical property of the physical universe defines a preferred position or direction in space is often viewed as defining an idealized model of the universe, like the assumptions that define idealized models of stars and galaxies. Unlike those assumptions, however, it *could* hold exactly. If the early universe is statistically uniform and isotropic, quantum indistinguishability implies that its *complete* description – and hence a complete description of all subsequent states – doesn't privilege any spatial position

5 Einstein, Albert, *Relativity, the Special and the General Theory*, 15th edition, , (Crown Publishers, New York, 1952).

6 Einstein, Albert, *The Meaning of Relativity*, 5th edition, (Princeton, NJ, Princeton University Press, 1953).

7 Einstein, Albert, *The Foundations of the General Theory of Relativity*, in *The Principle of Relativity*, (Methuen and Company 1923, reprinted by Dover Publications 1952).

8 Norton, John D., "The Hole Argument", *The Stanford Encyclopedia of Philosophy* (Fall 2015 Edition), Edward N. Zalta (ed.), URL = <https://plato.stanford.edu/archives/fall2015/entries/spacetime-holearg/>

or direction. I will refer to the hypothesis that it does hold exactly as the *strong cosmological principle*. This book argues that it is the missing element in a picture of the natural world that incorporates our strongly confirmed physical laws. It draws support from (at least) three arguments.

First, precise and extensive observations of the cosmic microwave background and of the spatial distribution and line-of-sight velocities of galaxies have so far produced no compelling evidence of deviations from statistical homogeneity and isotropy, now or in the past. Astronomical observations provide little support if any for the view that the cosmological principle is merely an approximation or an idealization, like the initial conditions that define models of stars and galaxies.

Second, the initial conditions that define the universe don't have the same function as those that define models of astronomical systems. A theory of stellar structure must apply to a range of stellar models because stars have a wide range of masses, chemical compositions, spins, and ages. But there's only one universe. The strong cosmological principle defines its simplest models. In that respect it is more like a law than an initial condition.

Finally, the strong cosmological principle accounts for what is sometimes called "Mach's principle": local inertial reference frames – frames in which Newton's and Maxwell's laws as well as their special-relativistic generalizations hold – are unaccelerated relative to a frame defined by the cosmic microwave background and the distribution and motions of distant galaxies. Einstein's theory of gravitation predicts this coincidence *provided the cosmological principle holds exactly*. Astronomical evidence supports this prediction. It indicates that local inertial reference frames are indeed unaccelerated relative to a coordinate system in which the cosmic microwave background is equally bright, on average, in all directions and the spatial distribution of galaxies is statistically homogeneous and isotropic. If the distribution of energy and momentum on cosmological scales were not statistically homogeneous and isotropic, there would be no preferred cosmological frame and hence no obvious explanation for the observed relation between local inertial frames and the frame

defined by the cosmic microwave background and the spatial distribution and line-of-sight velocities of galaxies. If the cosmic medium
was in thermal equilibrium at very early times, it lacked structure on
all macroscopic scales. Its state was maximally random. (As discussed
below, thermal equilibrium is a state of maximum randomness.) But
as the universe expanded, structure emerged. I have discussed one
possible scenario for this process elsewhere.[9] The cosmic distribution
of mass/energy became progressively less random, more orderly.

---

9   Layzer, David, *Cosmogenesis: The Growth of Order in the Universe*, (Oxford University Press, 1990).

IV

Entropy and Its Law

Heat as a Form of Energy

From the middle of the eighteenth century to the middle of the nineteenth century scientists who invented and used devices for measuring the heat released or absorbed in chemical reactions and in the compression and expansion of gas samples disagreed about the nature of what they were measuring. Some held that heat is a conserved substance, "caloric," analogous to mass. When a body gains caloric its temperature rises; when it gives up caloric its temperature falls. And in a heat engine, such as James Watt's steam engine, caloric does work on a movable piston – for example, by raising a weight – when the engine's working substance expands and heat "falls" from a higher to a lower temperature.　　　Newton, late in the seventeenth century, had championed a different view. It rested on what many scientists, then and until the opening years of the twentieth century, regarded as a speculative hypothesis: that the universe is made up of invisible, indivisible particles moving and interacting in otherwise empty space. Put forward by Leucippus in the fifth century BCE and elaborated by his pupil Democritus, the atomic hypothesis was made the basis of a comprehensive naturalistic philosophy by Epicurus (341 – 270 BCE). In the first century BCE it became the subject of a long poem, *On the Nature of Things*, by the Roman poet and philosopher Lucretius. Stephen Greenblatt, in *The Swerve*,[1] has argued that this widely read poem played an important role in the emergence of the modern world.

The atomic hypothesis suggested that a flow of heat into an otherwise undisturbed system, measured in an appropriate unit, produces an equal change in the sum of the kinetic energy associated with the motions of the system's hypothetical atoms and the potential energy associated with the forces the atoms exert on one another. Newton's laws of motion imply that the sum of these two kinds of energy is constant in time; an increase of one of them is accompanied by an equal decrease in the other. Inflows of heat increase an otherwise undisturbed system's internal energy; outflows diminish it. In the early 1840s Robert von Mayer (in Germany), James Joule (in England), and Ludwig Colding (in Denmark) independently advocated this hypothesis. In 1843 James Joule described experiments that lent it strong support. Using a calorimeter, he measured the quantity of heat generated by the viscous (internal frictional) dissipation of internal motions created by a paddle wheel immersed in water and driven by a descending weight. He found that the heat generated in this process, measured by a rise in temperature of the water, was consistently and accurately proportional to the quantity of mechanical energy that disappeared, measured by the decrease in the height of a weight whose descent drove the paddle wheel. And the constant of proportionality between the heat generated and the mechanical energy (in this case, gravitational potential energy) that disappeared always had the same value up to experimental error. In short, Joule concluded, heat has a fixed "mechanical equivalent." It is a kind of energy.

This proposition became the first of the two most basic laws of the new science of thermodynamics. It allowed scientists to extend the definition of energy from its original domain, mechanics, to include the domain of thermodynamics. They could now attribute a new property – internal energy –to an undisturbed macroscopic system that has relaxed into a macroscopically unchanging equilibrium state, and extend the scope of Newtonian mechanics' principle of conservation of energy, which up until then had applied only to systems whose internal motions don't generate heat, to any undisturbed system.

---

1   Greenblatt, Stephen , *The Swerve*, (New York: Norton and Company 2009).

The Second Law

The second of thermodynamics' two mains laws imposes a limitation on devices that, like the steam engine and the internal combustion engine, convert heat into mechanical energy. Its first version was a theorem stated and proved by the engineer and physicist Sadi Carnot (1796 - 1832) in an essay entitled "Reflections on the Motive Power of Heat" published in 1824. The essay opens:

> Everyone knows that heat can produce motion. That
> it possesses vast motive-power no one can doubt, in
> these days when the steam engine is everywhere so well
> known.

But despite its ubiquity in nature and its importance in industry,

> [t]he phenomenon of the production of motion by heat
> has not been considered from a sufficiently general point
> of view. We have considered it only in machines the na-
> ture and mode of action of which have not allowed us to
> take in the whole extent of application of which it is sus-
> ceptible. In such machines the phenomenon is, in a way,
> incomplete. It becomes difficult to recognize its princi-
> ples and study its laws. In order to consider in the most
> general way the principle of the production of motion by
> heat, it must be considered independently of any mech-
> anism or any particular agent. It is necessary to establish
> principles applicable not only to steam engines but to all
> imaginable heat engines, whatever the working substance
> and whatever the method by which it is operated.

To understand "the production of motion by heat," Carnot argues, it isn't enough to study heat engines experimentally. The principle underlying heat engines isn't an empirical generalization; it's an exact mathematical law. Carnot's essay lays bare this principle. It is known as Carnot's theorem.

A quarter of a century later, William Thomson (who became Lord Kelvin) and Rudolf Clausius independently deduced the second of thermodynamics' two main laws from a slightly but significantly emended version of Carnot's theorem. Thomson based his thermometer-independent definition of temperature (*thermodynamic temperature*) on Carnot's theorem, and Clausius based his definition of entropy and his derivation of the law of entropy change on Carnot's theorem and Thomson's definition.

In the following paragraphs I argue that the argument that led Clausius from Carnot's theorem to the conclusion that the entropy of the universe tends toward a maximum over-generalizes and over-extends a series of correct inferences. I conclude that his law of entropy change applies in an important but limited domain but doesn't apply to systems held together by the mutual gravitational attraction of their particles (self-gravitating systems). In particular, it doesn't apply to stars, galaxies, and the physical universe.

Neither Clausius's deduction of his law nor the argument that it doesn't apply to the physical universe involves mathematics beyond elementary algebra. Since that argument plays a pivotal role in the following discussion, I'll try to describe both Clausius's route to entropy and its law and my critique of the law's most general form in enough detail to allow readers with a modest mathematical background to make an informed judgment about the critique's validity.

Carnot's Theorem

Carnot's derivation of the theorem from which William Thomson and Rudolf Clausius deduced the second law of thermodynamics rests on two solid empirical notions – temperature and heat-gain (or heat-loss), both of which are unambiguously measurable (though in arbitrary units).[2] It also rests on two mistaken theoretical assumptions – that heat is a conserved substance and that it does work by "falling" from a higher to a lower temperature, just as water in a watermill does work when it falls from a higher to a lower elevation.

Carnot imagined an ideal cyclic heat engine, an engine that converts heat into work and whose working substance – the analogue of the steam in a steam engine – returns to its initial state at the end of each cycle. The engine's working substance is a gas enclosed in

---

2   Carnot's contemporaries measured a body's *temperature* by the volume of a sample of air or mercury in thermal contact with the body at a given pressure. Experimenters discovered that different "thermometric substances" yielded different but *inter-convertible* temperature measures. *Heat gain.* When heat flows into (or out of) a body, its temperature rises (or falls) – or else it undergoes a change of state, as when ice melts or water turns into steam. Experimenters discovered that different "calorimetric substances" yielded different but *interconvertible* measures of heat gain.

a cylindrical cavity bounded at one end by a movable piston. The working substance is intermittently in contact with two heat reservoirs: a source of heat, analogous to the boiler in a steam engine, and a sink of heat, analogous to the condenser in a heat engine. As the piston moves back and forth, the gas in the cylinder alternately expands and contracts. When it expands it does work on the outside world – for example, by raising a weight. When it contracts, the descending weight does work on the gas.

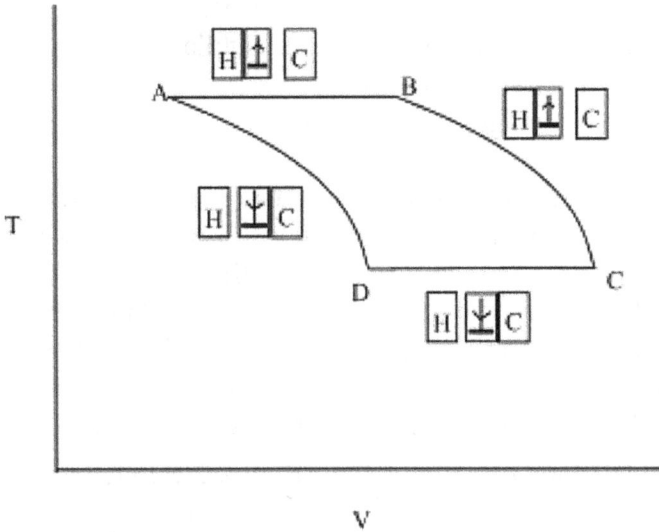

Figure 1   The Carnot Cycle

A point whose vertical coordinate is the working substance's temperature and whose horizontal coordinate is its volume represents the state of the working substance. At the beginning of a cycle the working substance is in the state represented by point A. Its temperature is equal to that of the hot reservoir (H), with which it remains in contact as its volume increases to the value represented by the horizontal coordinate of point B. As the working substance expands from state A to state B it does work on the piston and draws heat from the hot reservoir. Between states B and C the working substance neither gains nor loses heat but continues to expand and do work on the piston. Between states C and D it is in contact with and delivers heat to the cold reservoir (C) while the piston does work on it. In the final phase of the cycle, DA, the working substance neither gains nor loses heat as the piston does work on it. State D is chosen so that the fourth phase returns the working substance to its initial state. In the reverse cycle, ADCBA, the working substance expands, doing work on the piston while thermally isolated (AD), continues to expand while accepting heat from the cold reservoir (DC), contracts, having work done on it by the piston while thermally isolated (CB), and contracts further to its initial state while delivering heat to the hot reservoir (BA).

Carnot made his imaginary engine as efficient as possible by postulating not only that no mechanical energy is wasted by friction between the engine's moving parts but also that when heat flows between a heat reservoir and the working substance the two are at exactly the same temperature, thus ensuring that all of the heat withdrawn from the hot reservoir is transferred to the cold reservoir. He also assumed that when the working substance is thermally isolated it expands so slowly that at each moment it is in an equilibrium state characterized by definite values of its temperature and volume.

These idealizations not only make Carnot's ideal engine as efficient as possible. Crucially, they also make it reversible. Operating in its reverse mode, a Carnot engine acts as a refrigerator: it absorbs mechanical energy from the alternately falling and rising weight, extracts heat from the cold reservoir, and delivers heat to the hot reservoir.

Because he accepted the caloric hypothesis Carnot assumed that in the course of a cycle his ideal engine transfers all the heat it withdraws from the hot reservoir to the cold reservoir. Operating in its reverse mode, it transfers all the heat it withdraws from the cold reservoir to the hot reservoir.

Carnot defined the engine's efficiency as its mechanical-energy output during a cycle divided by the quantity of caloric transferred from the hot reservoir to the working substance (and eventually to the cold reservoir). He then asked and answered the question on which his fame rests: *Can two ideal heat engines operating between the same pair of heat reservoirs have different values of this ratio?*

Suppose this were possible. Let the more efficient engine transfer a quantity Q of caloric from the hot to the cold reservoir while doing a quantity of work W on its surroundings. We can then use the less efficient engine, operating in its reverse, or refrigerator, mode, to transfer the same amount of caloric Q from the cold reservoir back to the hot reservoir while doing a smaller quantity of work, W'. At the end of a cycle the composite engine is back in its initial state. Neither reservoir has gained or lost heat. But a quantity $W - W'$ of mechanical energy has appeared; the weight that rises during the expansion phase and falls during the compression phase is higher at the end of a cycle of the composite engine than it was at the beginning of the cycle. So if we could find two Carnot engines with different efficiencies we could build a perpetual-motion machine, a machine that creates mechanical energy out of nothing.

Carnot assumed that a perpetual-motion machine can't exist and concluded that all ideal heat engines operating between two heat reservoirs have the same efficiency. This conclusion is called *Carnot's theorem*.

Of course, the idealizations that define an ideal Carnot engine can't be realized in practice. Friction can't be entirely eliminated, heat doesn't flow between bodies at precisely the same temperature, and while a thermally insulated gas sample is expanding or contracting it is never in a state of equilibrium; it doesn't literally pass through a sequence of equilibrium states. Yet, as Carnot argued, actual heat engines can be made to resemble an ideal Carnot engine so closely that they would indeed create mechanical energy out of nothing if that were possible. So while Carnot's theorem rests on assumptions that can't be realized in practice, it is experimentally testable. And those assumptions allowed Carnot to discover the principle behind the motive power of heat – that all Carnot cycles operating between two heat reservoirs at given temperatures have the same efficiency.

From Carnot's Theorem to the Second Law

Carnot's theorem rests on the false assumption that heat is a conserved substance, like mass or energy in Newtonian physics. Having been convinced by Joule's experiments that heat and mechanical energy are interconvertible at a fixed rate of exchange, William Thomson and Rudolf Clausius independently set out to discover

what became of Carnot's argument when they replaced the false
caloric assumption with the newly established interpretation of heat
as a form of energy and the assumption that energy, in its new inclusive form, is conserved.

Carnot's conclusion that ideal Carnot engines working between
a given pair of heat reservoirs all have the same efficiency survived
this replacement. So did the idea behind Carnot's proof – hooking
up ideal engines with different efficiencies, one operating in its direct
mode, the other in its reverse mode. But Thomson and Clausius
could not now deduce from Carnot's argument that if two Carnot
engines had different efficiencies they could be coupled to make an
engine that created mechanical energy, for the First Law implies that
if in the course of a cycle the gas in the cylinder withdraws a quantity
of heat Q_2 from the hot reservoir and delivers a quantity of mechanical energy, or work, W to its environment, it must transfer a quantity
of heat $Q_1 = Q_2 - W$ to the cold reservoir; in the course of a cycle
the engine does an amount of work on its environment equal to the
difference between the heat it extracts from the hot reservoir and the
heat it delivers to the cold reservoir:

$$W = Q_2 - Q_1$$

We can continue to define an engine's efficiency as the ratio W/Q_2,
but that ratio now becomes

$$W/Q_2 = 1 - Q_1/Q_2,$$

Suppose now that two ideal Carnot engines working between a
given pair of heat reservoirs had different efficiencies. As in Carnot's
argument, we could then build a composite engine in which the first,
more efficient, engine operates in its direct mode and the second,
less efficient, engine in its reverse mode. The second engine accepts
a quantity of heat $Q_1{}^*$ from the cold reservoir, delivers a quantity of
heat $Q_2{}^*$ to the hot reservoir, and absorbs a quantity of mechanical
energy $Q_2{}^* - Q_1{}^*$.

Suppose we make $Q_1{}^*$ equal to Q_1. Then, because the second
engine is less efficient than the first engine, the preceding displayed
formula shows that $Q_2{}^*$ must be less than Q_2, so that $Q_2 - Q_2{}^* > 0$;
and the difference $Q_2{}^* - Q_1{}^*$, the work absorbed by the second, less
efficient engine during a cycle, is less than $W = Q_2 - Q_1$, the work

done by the first, more efficient engine. At the end of a cycle the composite engine has withdrawn a positive quantity of heat $Q_2 - Q_2{}^*$ from the hot reservoir and done an equal quantity of work on its surroundings. It *has converted heat drawn from a single reservoir into mechanical energy, leaving the world otherwise unchanged.* Alternatively, we can arrange matters so that the second engine absorbs all the mechanical energy the first engine delivers. *Then the composite engine extracts heat from the cold reservoir and delivers an equal quantity of heat to the hot reservoir.*

Thomson and Clausius independently surmised that both of these italicized statements describe impossible engines. This surmise quickly earned the status of a physical law, the second main law of the new science of thermodynamics. It has two equivalent forms:

> —*No cyclic engine can withdraw heat from a single source and convert it into mechanical energy, leaving the world otherwise unchanged.* In other words, a cyclic engine can't be perfectly efficient. It can't transform all the heat it withdraws from a heat source into mechanical energy; it needs a heat sink, to which it conveys some of the energy extracted from the hot reservoir.
> —*No cyclic engine can transfer heat from a cooler reservoir to a warmer reservoir, leaving the world otherwise unchanged.* In other words, a cyclic engine that transfers heat from a cooler reservoir to a warmer reservoir needs to be supplied with mechanical energy. Even if it is as efficient as possible, it must extract less heat from the cold reservoir than it deliver to the hot reservoir, the difference being equal to the mechanical energy that disappears in the course of a cycle.

Thomson and Clausius drew two remarkable inferences from the Second Law. Thomson used it to define temperature in a way that made it independent of thermometers. Clausius used the Second Law and Thomson's definition of temperature to define entropy and formulate the law of entropy change.

Thermodynamic Temperature

Before Thomson used Carnot's theorem to redefine temperature, scientists had defined a body's temperature as the reading of a thermometer that has been in thermal contact with the body long enough for both to reach equilibrium. They had formulated empirical laws that relate the volume and pressure of a gas sample to the temperature that would be measured by an ideal-gas thermometer – a thermometer that uses an ideal gas as the substance that expands or contracts when it absorbs or releases heat.

Since the work W done by a Carnot engine during a cycle is the difference $Q_2 - Q_1$ between the quantities of heat extracted from the hot reservoir and delivered to the cold reservoir, Carnot's theorem tells us that the engine's efficiency W/Q_2 depends only on the ratio Q_2/Q_1. This ratio takes an especially simple form if the working substance of a Carnot engine is an ideal gas. (The internal energy of an ideal gas sample depends only on the sample's temperature, and the sample's pressure, volume, and temperature are related by the "equation of state" $PV = nRT$, where R is a constant that has the same value for ideal gas samples of any chemical composition, and n is the number of gram-molecules (moles) in the sample.) From this definition of an ideal gas one can deduce that

$$Q_1/Q_2 = T_1/T_2 .$$

But Carnot's theorem shows that the efficiency of an ideal Carnot engine – and hence the ratio Q_1/Q_2 – doesn't depend on any property of the working substance. Thomson accordingly proposed, in 1848, that physicists *define* the ratio T_1/T_2 between the temperatures of two bodies as the ratio Q_1/Q_2 between the quantities of heat extracted from and delivered to the bodies when they act as (or are in thermal equilibrium with) the heat reservoirs in an ideal Carnot engine. The preceding displayed equation then shows that temperature ratios defined in this way coincide with temperature ratios measured by an ideal-gas thermometer. As Thomson explained,

> In the present state of physical science, therefore, a question of extreme interest arises: *Is there any principle on which an absolute thermometric scale can be founded?* It appears to me that Carnot's

theory of the motive power of heat enables us to give an affirmative answer.

The relation between motive power and heat, as established by Carnot, is such that *quantities of heat*, and *intervals of temperature*, are involved as the sole elements in the expression for the amount of mechanical effect to be obtained through the agency of heat; and since we have, independently, a definite system for the measurement of quantities of heat, we are thus furnished with a measure for intervals according to which absolute differences of temperature may be estimated.[3]

While the first law of thermodynamics makes heat independent of calorimeters, Thomson's definition, based on Carnot's theorem and the first law of thermodynamics, makes temperature ratios independent of thermometers.

Thomson's definition fixes the zero-point of the temperature scale but leaves the unit of temperature undefined. Thomson suggested that the unit of temperature be chosen to make the freezing and boiling temperatures of water at standard pressure be separated by one hundred units. The resulting unit of temperature, denoted by °K or just K, is called the Kelvin in his honor.

Entropy and the Law of Entropy Change

Thomson's definition of temperature paved the way for Rudolf Clausius's discovery, in 1854, of a previously unnoticed property of isolated macroscopic systems in thermal equilibrium. In 1865 he named this property *entropy*. Clausius deduced from the second law of thermodynamics that the entropy of an undisturbed system, unlike its energy and its mass, may change with time, and he proved that it never decreases. Later he extrapolated this law to the largest macroscopic system, the physical universe:

The entropy of the world tends toward a maximum.

After discussing Clausius's definition of entropy and his derivation of its law from the second law of thermodynamics, I'll argue that this extrapolation is invalid.

We can rewrite the equation that Thomson used to define temperature,

---

3   Thomson, William (Lord Kelvin), *Philosophical Magazine* October 1848.

$$Q_1/Q_2 = T_1/T_2,$$

as

$$Q_2/T_2 - Q_1/T_1 = 0.$$

This equation is suggestive. Suppose that the working substance in Carnot's imaginary engine absorbs a small quantity of heat, denoted by dQ, as it transitions between two neighboring equilibrium states. While the working substance is in contact with the hot reservoir at temperature T_2, these small heat transfers add up to Q_2; while it is in contact with the cold reservoir at temperature T_1, they add up to $-Q_1$. The preceding displayed equation tells us that in the course of a complete cycle the changes dQ/T add up to zero.

This conclusion suggests the question: Do the changes dQ/T add up to zero around *any* cyclic sequence of reversible transitions between equilibrium states of *any* macroscopic system? If the answer is yes, we can further infer that the sum of the small quantities dQ/T along any reversible path connecting any two equilibrium states, A and B, of the system, is the difference $S_B - S_A$ between the values of a property S of the system in these states.

To see why, suppose that S does indeed return to its initial value when the state of the system it refers to undergoes any reversible cyclic process. Suppose that one series of reversible changes, ABC, passes through an arbitrary state B and another series of reversible changes, CDA, brings the system back to its initial state A via another arbitrary state D. Since we are assuming that the quantities dQ/T sum to zero around any reversible path, they sum to zero along the path ABCDA. So the sum along the path CDA is the negative of the sum along the path ABC. But the sum along the path CDA is the negative of the sum along the reverse path ADC, because each summand in the sum of incremental changes dQ/T along ADC is the negative of the corresponding summand in the sum along CDA –when you reverse a heat inflow it becomes a heat outflow and vice versa. So the sum of the quantities dQ/T has the same value along the paths ABC and ADC – two arbitrarily selected paths connecting two given states, A and C. Which is what we set out to prove.

So if the quantity dQ/T sums to zero around any closed sequence of reversible changes, it represents a small change in a property of the system – a quantity that, like temperature, pressure, and internal energy, has the same value whenever the system is in an equilibrium state. Clausius named this property entropy. But does the quantity dQ/T sum to zero around any closed sequence of reversible changes?

To show that it does, Clausius invented a simple but ingenious argument.

Clausius's Argument

Recall that Carnot's proof of his theorem (that all ideal Carnot engines working between two heat reservoirs at given temperatures have the same efficiency) deploys two Carnot engines. One operates in its direct mode, withdrawing heat from the hot reservoir, depositing heat in the cold reservoir, and doing work on the piston. The other operates in its reverse mode, withdrawing heat from the cold reservoir, depositing heat in the hot reservoir, and having work done on it by the piston. Thomson's and Clausius's revised versions of Carnot's proof also use coupled pairs of Carnot engines, one member of each pair operating in its reverse mode. Clausius's proof that the quantity S returns to its initial value when the state of the macroscopic system it refers to undergoes any reversible cyclic process uses a generalized version of this idea.

Suppose a macroscopic system undergoes a sequence of n small, not necessarily reversible changes. At the kth change the system accepts a small quantity of heat dQ_k from a reservoir R_k at temperature T_k. dQ_k can be positive, negative, or zero. (So the word "accepts" doesn't have its everyday meaning in this context; but the new meaning will help us keep the signs straight.) The nth and final change brings the system back to its initial state, so the sequence of n changes constitutes a cycle. We can make n large enough so that the discrete sequence of changes approximates a continuous sequence as closely as we wish.

Clausius now imagines a collection of n ideal Carnot engines. The kth engine works between reservoir R_k at temperature T_k – the temperature of the system during the kth change – and a reservoir R_0 used by all n Carnot engines. The kth engine replaces the heat drawn by the system from the reservoir R_k by delivering an equal amount of heat dQ_k to R_k and it withdraws a quantity of heat dQ_{k0} from the common reservoir R_0. Thomson and Clausius showed in their revised version of Carnot's proof that

$$dQ_k/T_k = dQ_{k0}/T_0 \quad (k = 1, 2, ...,n)$$

Summing these n equations, we get:

$$\Sigma_k \, dQ_k/T_k = (\Sigma_k dQ_{k0})/ \, T_0 \, ,$$

where the symbol Σ_k indicates that the quantity on its right is to be summed over all values of the index k between 1 and n. The sum Σ_k dQ_{k0} represents the total quantity of heat that has been withdrawn from the reservoir R0 in the course of a cycle. Call this quantity Q0.

At the end of a cycle the reservoirs with labels from 1 to n are back in their initial states, and so is the system. But a quantity of heat Q0 has been withdrawn from the reservoir R0. Suppose Q0 is positive. Since energy is conserved, the cyclic process just described must produce an equal quantity of mechanical energy. But the Second Law says this can't happen; heat drawn from a single source can't be transformed entirely into mechanical energy. So Q0 must be non-positive.

The last displayed equation then tells us that the sum $\Sigma k \, dQ_k/T_k$ is also non-positive:

$$\Sigma_k \, dQ_k/T_k \leq 0 \text{ for any cyclic process} \quad (1)$$

If the cycle is reversible, the preceding equation still holds if we reverse the sign of each of the quantities dQ_k. But this is impossible unless the sum $\Sigma_k \, dQ_k/T_k$ equals zero:

$$\Sigma_k \, dQ_k/T_k = 0 \text{ for a reversible cyclic process} \quad (2)$$

Earlier we saw that if S is a quantity that changes by an amount dQ/T when the system it refers to reversibly absorbs a quantity of heat dQ, and if S returns to its initial value when the system undergoes any reversible cyclic process, then the difference $S_C - S_A$ doesn't depend on the path that connects A to C. Equation (2) says that S does indeed returns to its initial value when the system undergoes any reversible cyclic process. So S is indeed a physical property of the system, the property Clausius named entropy.

The preceding argument defines the entropy S of a system in thermal equilibrium through the amount by which S changes when the system passes from one equilibrium state to another via a series of reversible changes. Consequently, S is defined only up to an additive constant, which we may identify with the entropy of the system in an arbitrary reference state.

Next suppose that an isolated, or undisturbed, macroscopic system evolves from equilibrium state A to equilibrium state C via an irreversible process. Because the system is isolated, it doesn't interact with the outside world; no heat flows into or out of it and no work is done on or by it.

For example, imagine a box divided into two compartments by a removable partition. Suppose that initially, one compartment is filled with air at temperature T_2, the other with air at a lower temperature T_1, and assume (as can always be arranged) there's no pressure difference between the compartments. After the partition is removed, the sample settles into an equilibrium state at a single uniform temperature between T_1 and T_2. If during this process a quantity of heat Q flows from the set of molecules that make up the subsample that was initially at temperature T_2 to the set that make up the subsample that was initially at temperature T_1, the entropy of the whole sample changes by an amount $(Q/T_2 - Q/T_1)$. In this example the entropy of the system increases (because $T_2 > T_1$). We want to show that the entropy of any isolated, or undisturbed, system either increases or doesn't change when it passes from one state of thermal equilibrium (in which all its macroscopic properties have definite, unchanging values) to another.

Call the initial equilibrium state A and the final equilibrium state C. Now construct a cycle by adding a *reversible* sequence of changes CDA, in which heat can flow into or out of the system and work can be done on or by it. (Any two equilibrium states can be joined by infinitely many reversible sequences of reversible changes, represented by smooth curves joining the points that represent the two states.) Applying equation (1) to the cycle $ACDA$ gives:

$$\Sigma_{AC} \, dQ_k/T_k + \Sigma_{CDA} \, dQ_k/T_k \leq 0$$

Because no heat flows into or out of the system during the irreversible leg AC, the first sum vanishes. The second sum is $S_A - S_C$, the entropy change between C and A. So

$$S_C \geq S_A \qquad\qquad (3)$$

When a closed, or isolated, system evolves between two equilibrium states, its entropy cannot decrease; it must either increase or stay the same.

This is the law of entropy change. Clausius deduced it, in the way I've described, from the Second Law of thermodynamics. His argument shows that if the entropy of an isolated system were to decrease, we could build a device – the device described in Clausius's proof – that converts all the heat drawn from a single source into mechanical energy.

The law doesn't say that the entropy of an isolated system must increase during an irreversible change in the system's state; it merely says that it cannot decrease. For example, if an isolated system consists of two subsystems with initially different temperatures, the law tells us that heat can flow only from the warmer to the cooler subsystem, because heat flow in the opposite direction would decrease the system's entropy. But the law doesn't tell us that heat *must* flow between the subsystems. As in our earlier example, the two subsystems might be separated by a membrane that doesn't conduct heat.

A separate physical law governs heat flow. It contains a parameter that characterizes the thermal conductivity of a medium at each point and in every direction. If you change the sign of the time coordinate in this law, and hence the direction in which time increases, you get a different law – one that predicts that heat flows *up* a temperature gradient and thus violates the law of entropy change.

More generally, separate macroscopic laws govern the processes through which an isolated system evolves from one equilibrium state to another through a sequence of irreversible processes. The law of entropy change constrains the laws that govern these processes, but doesn't, on its own, require that irreversible processes cause the entropy of an isolated system not initially in thermal equilibrium to increase.

Defining Entropy for Systems Not in Thermal Equilibrium

What is the scope of Clausius's definition of entropy? So far we've defined entropy for homogeneous systems, such as uniform gas samples, in thermal equilibrium. The preceding proof of the law of entropy non-decreases applies to two states, A and C, of such a system connected by an irreversible path. Now consider an undisturbed nonuniform gas sample. Can we assign entropies to its states between A and C and apply the law of entropy non-decrease to them? For example, can we assign entropy to an isolated gas sample whose temperature varies smoothly from point to point?

Clausius defined the entropy S of a system in thermal equilibrium through its changes between neighboring equilibrium states. The First Law of thermodynamics tells us that if the volume V of a gas sample enclosed in a cylinder fitted with a movable piston changes by an amount dV and a positive or negative quantity of heat dQ flows into the sample, its internal energy U changes by an amount dU given by:

$$dQ = dU + P\,dV.$$

Here P is the pressure the sample exerts on the piston. If the piston has area A and travels a short distance dx, it does work $PA\,dx$ or $P\,dV$. So the preceding equation says that the work done by the sample when its volume changes must come from either the sample's internal energy or from heat that flows into the sample or from a combination of the two. The preceding equation shows that unlike dU and dV, dQ isn't the change in a quantity that characterizes the state of the gas sample; although heat is a form of energy and energy is conserved, a gas sample doesn't hold heat in a separate account. It does, though, hold entropy in a separate account. Clausius's theorem enables us to replace dQ by TdS, where dS, like dU, dV, and dT, denotes a small change in a quantity that characterizes the state of the gas sample. The preceding equation then takes the form

$$TdS = dU + PdV \qquad \text{or}$$
$$dS = (dU/T) + (P/T)dV$$

Suppose now that we can draw imaginary surfaces separating the region occupied by an inhomogeneous gas sample into cells in each of which the temperature and pressure are (nearly) uniform. We will be able to do this if the gas sample isn't too chaotic. But if the entropy of the sample is to equal the sum of the entropies of our imaginary cells, two more conditions must be satisfied: The sample's energy U must be *additive*: it must equal the sum of the energies of the parts. And work performed by the system must be equal to the sum of the amounts of work PdV performed by the parts. Then, and only then, can we equate the change dU in the sample's energy to the sum of the energy changes of the parts. Enrico Fermi emphasized this condition in his lectures on thermodynamics:

> The entropy of a system composed of several parts is very often equal to the sum of the entropies of all the parts. This is true if the energy of the system is the sum of the energies of all the parts

and if the work performed by the system is equal to the sum of the amounts of work performed by all the parts. Notice that these conditions are not quite obvious and that in some cases they may not be fulfilled. Thus, for example, in the case of a system composed of two homogeneous substance, it will be possible to express the energy as the sum of the energies of the two substances only if we can neglect the surface energy of the two substances where they are in contact. The surface energy can generally be neglected only if the two substances are not very finely divided; otherwise it can play a considerable role. [4]

Energy is usually additive under laboratory conditions. Parcels of air in the atmosphere, whose energies include a height-dependent gravitational potential energy, also satisfy the additivity condition, as do parcels of gas in the interiors of stars. But the Earth as a whole and the Sun as a whole don't satisfy the condition, because the gravitational potential energy of a system held together by the mutual gravitational attraction of its parts isn't a sum of contributions each of which depends only on local conditions. This is a consequence of what physicists call the "long-range" character of Newtonian gravitational force – the fact that it decreases only with the inverse square of the separation between mutually attracting particle. This makes gravitational potential energy a nonlocal property, in contrast to the energies associated with the short-range forces that come into play when two gas atoms or molecules collide, or the binding energies of atoms in gas molecules or of atoms and molecules in crystals. So Clausius's definition of entropy doesn't apply to planets, stars, galaxies, or other self-gravitating systems. Consequently, it doesn't apply to the universe.

An isolated macroscopic gas sample settles into thermal equilibrium – a macroscopically quiescent state of uniform temperature. Could a self-gravitating system be in such a state? It could be in a state of *mechanical* equilibrium, in which the weight of every small parcel of gas balances the pressure force the surrounding gas exerts on the parcel's boundary. But a calculation based on Archimedes's law of hydrostatic equilibrium shows that if the system also had uniform temperature – if it were in thermal as well as mechanical equilibrium – it would extend to infinity and have infinite mass. *A finite self-gravitating system can't evolve toward a state of uniform temperature – a state without internal heat flow.*

---

4   Fermi, Enrico, *Thermodynamics*, Prentice-Hall, New York, 1937, p. 53.

In a star, energy is liberated by thermonuclear reactions in the deep interior and escapes from the surface layers in the form of light and neutrinos. As a result, the star's structure and its energy slowly change. In a protostar whose central temperature is not yet high enough for thermonuclear reactions to liberate energy, the loss of energy by radiation from the surface layers is uncompensated, so the protostar's energy decreases. This has a surprising consequence. The *virial theorem* (like entropy and its law, discovered and named by Rudolf Clausius) implies that when a self-gravitating system of particles is in mechanical equilibrium, its gravitational energy, a negative quantity, is equal in magnitude to twice the system's thermal energy (the combined kinetic energy of its particles). The sum of the system's gravitational energy and its thermal energy is consequently a negative quantity equal in magnitude to its thermal energy (or half its gravitational energy). So when the system *loses* energy, its thermal energy, and hence its mean temperature, *increase*. Thus a self-gravitating system in mechanical equilibrium has negative heat capacity (the ratio between a small inflow of heat – in the present example a negative inflow – and the consequent small increase in temperature).   In contrast, the second law of thermodynamics predicts (and experiments confirm) that when a system (to which the law applies) loses energy, its temperature drops. Of course, as we've just seen, the domain of the Second Law doesn't include self-gravitating systems – and for basically the same reason that a self-gravitating gas mass grows hotter as it radiates away energy: gravitational potential energy is non-local and hence non-additive.

Rudolf Clausius's discovery of entropy and his reformulation of the Second Law as the law of entropy change enabled him to recast thermodynamics as a deductive mathematical science, an adjunct to Newtonian particle physics and Newtonian fluid mechanics. That science has proved to be enormously useful, and its predictions have been uniformly successful in a very wide domain. But, as we've seen, that domain is limited to macroscopic systems whose structure is sufficiently smooth and whose energy can be expressed as a sum of the energies of nearly uniform subregions. As long as "the entropy of the universe" remains undefined, we can assign no meaning to Clausius's statement that the entropy of the universe tends toward a maximum.

V
Atomism

Thermodynamics' two main laws are generalizations about the outcomes of possible experiments. The first law says that in any experiment in which heat appears in a closed (or isolated or undisturbed) system a strictly proportional quantity of mechanical energy, such as gravitational potential energy, disappears. The second law states that it is impossible to build a device whose only effect is to transfer heat from a cooler to a warmer body; or, equivalently, to build a device whose only effect is to transform heat drawn from a single heat reservoir into mechanical energy. Some of Clausius's contemporaries succeeded in constructing a theory based not on these empirical generalizations but on Newton's laws of motion together with an assumption that, in the mid-nineteenth century, seemed far less secure than thermodynamics' two empirical laws: that matter consists of invisible and indivisible particles in motion. Not until the early 1900s did that assumption, the atomic hypothesis, become the atomic fact (as the physicist Richard Feynman put it in his *Lectures on Physics*).

Atomic theories of macroscopic systems and processes have a statistical character; they are formulated in the mathematical language of probability theory. Ludwig Boltzmann showed that the statistical description of an isolated sample of an ideal gas has a counterpart to entropy. As we'll discuss, this counterpart, sometimes called *statistical entropy*, is well defined in a much broader domain than Clausius's entropy. It is a measure of the randomness inherent in a statistical description. So for the sake of clarity, I'll usually call it *randomness* rather than statistical entropy.

While probability theory itself is uncontroversial, its interpretation in statistical theories based on the atomic hypothesis raises a question that goes to the heart of the problem of free will versus determinism: *Do the probabilities that figure in these theories represent incomplete knowledge? Or do they represent a kind of objective indeterminacy distinct from quantum indeterminacy?* To answer this question we'll need to take a closer look at atomism and statistical physics.

The Roots of Atomism

In the fifth century BCE Leucippus and his pupil Democritus (460 – 370 BCE) conjectured that the world consists of indivisible particles, or atoms, moving about in otherwise empty space. Atoms came in a variety of sizes and shapes. Some had hooks or barbs on their surfaces, allowing them to form molecules; others, such as water atoms, were smooth. Atoms of fire and soul were small and round and moved at great speeds.

Epicurus (341–270 BCE) based his philosophy on this thoroughly materialistic picture of the world. But he amended it in one important way. In his ethical philosophy Epicurus assumes that we are free to shape our conduct. But how does freedom fit into a world in which our thoughts and actions are determined by invisible atoms and their motions? Epicurus suggested that there is an element of randomness in atomic motions. Atoms occasionally swerve from their paths, and these swerves have an irreducibly random character. This assumption allowed Epicurus to reconcile his materialistic picture of the world with an ethical philosophy based on human freedom.

How did the atomists come up with their picture of the world? In his *History of Western Philosophy* Bertrand Russell suggested that the atomic hypothesis was a lucky guess:

> By good luck, the atomists hit on a hypothesis for which, more than two thousand years later, some evidence was found, but their belief, in their day, was nonetheless destitute of any solid foundation.[1]

This judgment rests on a widely accepted view of the "scientific method" in the physical sciences:

---

1   Russell, Bertrand, *History of Western Philosophy*, (New York, Simon & Schuster, 1967) p. 68.

> Scientists do experiments and make observations. These
> may suggest hypotheses that have testable implications.
> Further experiment or observations then confirm or
> disconfirm these hypotheses. Those that survive become
> physical laws.

The same view of the scientific method underlies the definition of physical laws in the *Oxford English Dictionary*, 3d edition: Physical laws are statements "inferred from particular facts, applicable to a defined group or class of phenomena, and expressible by the statement that a particular phenomenon always occurs if certain conditions be present." Definitions like these reflect a view of physical laws that was popular (though not uncontroversial) among physical scientists in the late nineteenth century. Its most famous and influential advocate was the physicist and philosopher of physics Ernst Mach. Mach held that physicists should abjure metaphysical notions, such as Newton's "absolute" space and "absolute" time; they should rely entirely on objective experimental and observational facts. Einstein tells us that Mach's elaboration of this view in *The Science of Mechanics* helped motivate his own search for a theory that would relate the structure of spacetime to its contents.

But Mach went further. He held that physical laws are nothing more than economical summaries of experimental and observational data. So he rejected theoretical constructs that went beyond the data. These constructs included atoms and Einstein's principle of special relativity. During the opening years of the twentieth century, Mach's view of physical laws fell out of favor. One group of experiments, which included experiments designed to test Einstein's predictions about the random motions of microscopic particles suspended in a liquid (Brownian motion), showed that atoms actually exist. Another group of experiments confirmed special relativity's predictions about phenomena involving particles traveling at nearly the speed of light. Though physicists continued to insist, and still insist, that scientific hypotheses stand or fall by their testable predictions, special relativity and the experimental vindication of the atomic hypothesis showed that not all scientific hypotheses are empirical generalizations. Could Leucippus and Democritus have viewed atomism as a scientific hypothesis? Would they even have understood the notion of a scientific hypothesis?

About Leucippus we know only that he was Democritus's mentor. Democritus, however, was famous in his day as a mathematician as well as a philosopher. He wrote treatises on number theory, geometry, and astronomy. And he was no run-of-the-mill mathematician (if such people exist). According to the historian of Greek mathematics Thomas Heath, Democritus "already had the idea of a solid being the sum of an infinite number of parallel planes, or indefinitely thin laminae [plates], indefinitely near together; a most important anticipation of the same thought which led to such fruitful results in Archimedes." The thought in question is the idea underlying the integral calculus, an early version of which Archimedes used to calculate the area of a segment of a parabola and the surface area and volume of a sphere. Two centuries before Archimedes, Democritus used the same idea to prove that the volume of a circular cone is one-third the product of its height and the area of its base, and that the volume of a pyramid is likewise one-third the product of its height and the area of its base. So Democritus was, to say the least, an exceptionally creative and insightful mathematician.

Nowadays many mathematicians specialize in subjects remote from the concerns of the natural sciences. But Greek mathematics, physics, and astronomy were closely interwoven strands of a single project. Heath tells us that these subjects "were born together at the beginning of the sixth century."[2] Collections of geometric and arithmetic facts, rules, and algorithms had existed for millenniums in Egypt and Mesopotamia before Thales of Miletus (about 624 – 547 BCE), after visiting Babylon and Egypt and acquainting himself with some of their mathematical treasures, invented a new kind of mathematics: He proved the first theorems.

A theorem is a statement about mathematical objects and relations accompanied by a proof. Written out in full, a proof is a chain of logical deductions that connects the statement to be proved to a small fixed collection of unproved statements, or axioms, and a small fixed collection of terms, such as *point* and *line* in plane geometry, that are implicitly defined by the axioms that mention them. Logical deductions preserve truth: if the premises of a deduction are true, then so is its conclusion. So if geometry's axioms are true, then so are its theorems. Greek mathematicians and their successors until well into the nineteenth century regarded the axioms of number theory

---

2   Heath, T. L, *A History of Greek Mathematics*. (Cambridge University Press, 2013).

and geometry as self-evident truths. By contrast, many *theorems* about numbers and geometric objects were far from self-evident. Some were amazing. For example, Greek mathematicians discovered and proved that the lines connecting the vertices of any triangle to the midpoints of the opposite sides intersect in a single point, which divides each of these lines into two segments one of which is half the length of the other.

Again, Pythagoras (about 570 - 490 BCE) or one of his followers discovered that *geometric* atomism is false: a side and a diagonal of a square have no common measure; it isn't possible to choose a unit of length so that the lengths of a side and a diagonal are both whole-number multiples of that unit. And since this claim could be proved, it had to be true.

Mathematics is a creative enterprise. Theorems and their proofs need to be invented. Mathematicians imagine properties of numbers or geometric objects that might be true; then they imagine possible routes to a proof. For its first practitioners the method of conjecture and proof was a way of arriving at hidden truths about the world, more reliable than divine revelation or authority. And its successes seemed to reveal that numbers and geometric forms underpin reality. Mathematics was the language of the new sciences of physics and astronomy. Even the parts of Greek philosophy that were not explicitly mathematical aspired to the logical rigor epitomized by mathematical demonstrations. Plato held that the study of mathematics was a necessary propaedeutic to the study of philosophy. The motto "Let no one unversed in geometry enter here" was inscribed over the door of his Academy.

Greek mathematicians didn't draw a sharp distinction, as we do today, between pure mathematics and mathematical physics. They recognized that mathematics deals with idealized objects – that dimensionless points, perfectly straight lines, and perfectly smooth planes exist nowhere in the world of experience. But mathematical discoveries and the method of conjecture and proof seemed to reveal hidden and often surprising truths not only about idealized mathematical objects and relations but also about the physical world. Theorems about plane figures and solids are also testable propositions about their inexact real-world counterparts. And experience invariably supported these propositions to within the accuracy of the measurements. Mathematics as a deductive science (rather than as a set of rules, formulas, and algorithms) was seen as a new and powerful tool in humankind's struggle to understand and control nature, allowing its practitioners to glimpse a hidden reality that undergirds experienced reality but can be described only in mathematical language.

Thales, reputed to have been the first mathematician, devised a method for finding the distance of a ship at sea and a method for calculating the heights of the pyramids, both based, presumably, on the theorem that pairs of corresponding sides of similar triangles have the same ratio. Both methods also rely on identifying light rays or lines of sight with the straight lines of theoretical geometry. By combining geometry with observation, Greek astronomers conjectured and then proved that the Earth and the Moon are spheres. They also arrived at the modern explanation of solar and lunar eclipses. In the fourth century BCE, Heraclides suggested that the daily motions of the stars are produced by the Earth's rotation (rather than by the rotation of a solid Earth-centered sphere in which the stars are embedded). A century later Aristarchus anticipated Copernicus's model of the planetary system, in which the Moon is a satellite of Earth, Earth is a planet, and the planets are satellites of the Sun. Aristarchus also invented and applied a method of deducing the ratio of the Sun's and the Moon's distances from observations of their directions when the Moon is exactly half full. His account marks the debut of trigonometric ratios such as sines and tangents.

Archimedes took over the method and style of Euclid's *Elements* (composed around 300 BCE) and extended its vocabulary to include the notions of mass, force, and mechanical equilibrium. He invented the notion of center of mass, discovered the principle of the lever (objects at opposite ends of a weightless beam balance if the ratio of their weights is equal to the reciprocal of the ratio of the distances of their centers of mass from the fulcrum), and founded the science of hydrostatics, which includes the proposition known as Archimedes' principle: An object wholly or partly submerged in water experiences an upward, or buoyant, force equal in magnitude to the weight of the water it displaces; so the weight of a submerged object is equal to its weight in air minus the weight of an equal volume of water; if the difference is negative, the object experiences a net upward force.

Archimedes's *The Method of Mechanical Theorems* used his method of infinitesimals, along with what we'd now call physical concepts – the notion of center of mass and the principle of the lever – to calculate areas and volumes of *geometric* objects. For Archimedes, number theory, geometry, and mechanics were interlocking parts of a single mathematical structure. Archimedes was famous in the ancient world for putting the physical principles he had discovered to practical uses. His inventions include the screw pump (for bailing ships), block-and-tackle pulley systems, and a mechanical model of the solar system. He also invented devices that were used to defend his native Syracuse, in Sicily, from the invading Romans: a powerful catapult, a device for lifting ships out of the water and then dropping them (Archimedes's claw), and (perhaps) arrays of either lenses or parabolic mirrors for focusing the Sun's rays on an invader's ships and setting them on fire.

To return to the atomic hypothesis, we can perhaps think of it, in the context of Greek mathematics and natural science, as the response to a question that is both philosophical and scientific. One way to state the question is to ask what would happen if you split a drop of water in two, then split each of the resulting drops in two, and kept on in this way. Leucippus and Democritus might perhaps have reasoned as follows.

Either the process of splitting will go on forever or it will come to an end. If every substance is indefinitely divisible into smaller and smaller parts with the same physical properties, the only possible answer to the question "Why do different substances have different properties?" is "They just do." If the splitting process eventually ends, we have a choice. Either the process ends with a large collection of indivisible water droplets or with a large collection of objects that aren't water droplets. The first possibility doesn't help us to understand why alcohol differs from water. If, as I'm assuming, Leucippus and Democritus were seeking to understand such differences, they would have opted for the second possibility – that if you continue to split a drop of water you will eventually arrive at particles whose properties don't coincide with *but serve to explain* the properties of water. In the historical context I have sketched, these would have been objects describable in the vocabulary of Euclidean solid geometry. If you ask the right question – a question that would have arisen quite naturally in the context of Greek mathematics, physics, and natural philosophy – the atomic hypothesis seems almost inevitable.

We can interpret the question "What happens when you keep splitting a bit of matter into smaller and smaller parts?" as either a philosophical or a scientific question, depending on the kind of answer we're looking for. Physicists ask questions whose answers can be expressed in the vocabulary of mathematics, even if they have to invent new mathematics to do so. Biologists ask questions that fit into a conceptual framework based in part on the premises that all living organisms are genetically related and have evolved from a single ancestral population. Empiricists ask questions that can be answered by an appeal to sensations and perceptions; metaphysicians, by an appeal to basic metaphysical principles, such as Gottfried Wilhelm Leibniz's principles of sufficient reason (everything has a reason) and identity of indiscernibles (distinct objects can't have identical properties).

Atomism Becomes a Testable Hypothesis

The atomic hypothesis had immediate explanatory value. It supplied a starting point for an explanation of why water becomes a solid at low temperatures and becomes a vapor at high temperatures. It explained why you can mix wine and water in various proportions to make homogeneous liquids with experimentally distinguishable properties. It explained why a sample of brine exposed to air eventually turns into salt crystals. Newton, with not much more *direct* evidence than Democritus had had, was a convinced atomist. To quote the historian of the physical sciences Alan E. Shapiro:

> The corpuscular theory of matter was thus for Newton not a hypothesis but a demonstrated principle established with as much certainty as the existence of God or the theory of gravitation. He cites two principal sorts of evidence in its support: various substances penetrate the pores [empty spaces between atoms] of bodies, like water into vegetable and animal matter, and quicksilver [mercury] into metals; and transparency, which shows that light passes through the pores of a great variety of bodies [such as thin layers of gold][3]

While Newton considered his view that matter consists of atoms to be "a demonstrated principle" on a level with his law of gravitation,

---

    3   Shapiro, A.E., *Fits, Passions and Paroxysms: Physics, Method and Chemistry and Newton's Theories of Colored Bodies and Fits of Easy Reflection.* (Cambridge University Press, 1993).

he considered another of his strongly held views – that light rays are streams of particles – to be a hypothesis in need of experimental support – which his own experiments consistently *failed* to provide.

Newton's view of atoms and their role in observable phenomena differed from Democritus's in a crucial respect: Newton assumed that the motions and interactions of atoms obey his three laws of motion. Thus enriched, Democritus's picture of the physical world became a testable scientific hypothesis. It not only offered qualitative explanations of some phenomena but could also serve as the basis for testable predictions about the outcomes of precise measurements. Prominent among the empirical rules Newton hoped the atomic hypothesis could explain was Boyle's law.

In 1659 Robert Boyle, assisted by Robert Hooke, built an improved version of Otto von Guericke's air pump, and with its help began a series of experiments on what Boyle called "the spring of the air" – the fact that a finite sample of air resists compression and also expands to fill any airtight enclosure. Their experiments supported the quantitative relation between the pressure and the volume of an enclosed gas sample (at a given temperature) that is now called Boyle's law: the pressure, or force per unit area, the sample exerts on the walls of its container is inversely proportional to the sample's volume.

To account for this remarkably simple rule, Newton proposed that air atoms repel one another. He showed that this hypothesis would yield Boyle's law if (a) each gas atom repelled only its nearest neighbor and (b) the repulsive force was inversely proportional to the distance between the two atoms. "But whether elastic fluids do really consist of particles so repelling each other," he wrote in the *Principia*, "is a physical question" – that is, a question to be settled by experiment.

And experiment did settle it. Although Newton's model accounted for Boyle's law, another of its predictions contradicted an empirical law established by Guillaume Amontons between 1700 and 1702: *If an air sample's volume is held constant, its pressure increases with its temperature.* Newton's model predicts instead that the force a gas sample exerts on the walls of its container depends only on the average number of atoms per unit volume and not on the temperature.

In his *Hydrodynamica*, published in 1738, Daniel Bernoulli proposed an atomic model of gases that not only accounted much more simply for Boyle's law than Newton's model but also predicted Amontons' law and other, as-yet-undiscovered, empirical laws.

Like Newton, Bernoulli postulated that in a uniform gas sample the number of gas particles per unit volume has the same value everywhere. But instead of postulating that the particles of a quiescent gas sample are themselves motionless, as Newton had done, he postulated that they have large random velocities. He imagined that each particle travels at constant speed in a fixed direction until it collides with another particle or with a wall, after which it sets off in a new direction. He assumed that the duration of a collision is much shorter than the time interval between collisions, as it is in collisions between billiard balls. (Bernoulli's successors assumed further that particle collisions are elastic: the combined momentum of a pair of colliding particles is the same before and after a collision and so is their combined kinetic energy.)

Newton had attributed the expansive tendency of a gas sample to a repulsive force between the gas's constituent particles. In Bernoulli's model the fact that a gas sample expands to fill a box of any size follows directly from Newton's first law of motion: a particle on which no force acts continues to move in a straight line with constant speed; collisions with other particles deflect a gas particle but can't confine it to a finite region.

Appealing to Newton's second and third laws of motion, Bernoulli equated the force an enclosed gas sample exerts on a wall of its container to the rate at which gas particles transfer momentum (mass times velocity) to the wall when they strike it and rebound. He assumed that such encounters are elastic, conserving momentum. The force per unit area, or pressure, is then proportional to the number of particles per unit volume and to two-thirds the average value of a particle's kinetic energy (half the product of its mass and the square of its velocity). So Bernoulli's model predicts the following simple relation between the pressure, P, the number of gas particles per unit volume, n, and the average kinetic energy of a particle, $\langle \tfrac{1}{2} mv^2 \rangle$:

$$P = \tfrac{2}{3}\, n \,\langle \tfrac{1}{2}\, mv^2 \rangle \,\langle E \rangle \,\langle E \rangle$$

If we now *assume* that the temperature T of a gas is some constant multiple of the mean kinetic energy of the gas molecules, Bernoulli's formula becomes

$$P = constant \times nT = constant \times (N/V)T,$$

where N is the number of particles in a sample, V is the sample's volume, and n is the number of particles per unit volume. This relation includes not only Boyle's law and Amontons' law (which says that the pressure in a gas sample whose volume is held constant increases with its temperature) but also Charles's law (1780), which says that at constant pressure the temperature of an ideal-gas sample is proportional to its volume. Moreover, because the displayed equation doesn't depend on any property of the gas particles, it contains Amedeo Avogadro's hypothesis (1811): *Equal volumes of different ideal gases at the same temperature and the same pressure contain equal numbers of particles (N in the displayed equation).*

We can't call the preceding argument a *derivation* of the gas laws, because it merely *assumes* that the average kinetic energy of the gas particles is proportional to the gas sample's temperature, as defined by Thomson through Carnot's theorem. It doesn't explain why. As we'll see, a full derivation of the ideal-gas law didn't appear until over a century after Bernoulli proposed his model.

Atoms and Molecules

During the eighteenth century chemists devised experimental methods for distinguishing homogeneous "substances" from mixtures, and for distinguishing "elements" – substances that couldn't be broken down into simpler substances –from "compounds." They perfected the chemical balance and used it to discover that elements combine to form compounds in fixed proportions by weight. They also discovered that weight (or mass) is conserved when a compound is formed or when it is separated into elements.

During the first decade of the nineteenth century John Dalton (1766 – 1844) realized that simple assumptions about the atomic composition of elements and compounds could bring greater order to the wealth of experimental data chemists had built up during the preceding century. He postulated that every chemical element consists of identical, unchanging atoms that differ from the atoms of other elements, and that chemical compounds consist of identical molecules made up of atoms of the compound's constituent elements.

These postulates immediately accounted for a striking regularity in the data. Sometimes two elements combine to form different compounds. For example, carbon and oxygen combine to form the gases we now call carbon monoxide and carbon dioxide. Experiments showed that if samples of the two compounds contain equal masses of carbon, the mass of oxygen in the heavier sample is twice the mass of oxygen in the lighter sample. To take another stock example, arsenic and oxygen likewise form two compounds. If samples of these compounds contain equal masses of arsenic the ratio between the masses of oxygen in the two samples is exactly 5/3 – again a ratio of two small whole numbers. These examples suggest, and experiment confirms, the following rule: If element #1 forms two (or more) different compounds with element #2, then samples that contain the same mass of element #1 contain masses of element #2 that are in the ratio(s) of small whole numbers. Before Dalton, this finding seemed inexplicable. Dalton's postulates make the explanation obvious: *Every molecule made up of atoms of different kinds contains a small integral number of each kind.*

Dalton's axioms by themselves didn't allow chemists to predict the proportions of elements in every compound. In the preceding examples you might guess, correctly, that the two oxides of carbon are, in modern notation, CO and CO_2, but you might also guess, incorrectly, that the oxides of arsenic contain a single arsenic atom (in fact they contain two). Dalton suggested that a "rule of simplicity" should guide the choice of chemical formulas for compounds. But efforts to follow this rule led to inconsistencies. Then in 1808, the same year in which Dalton's *A New System of Chemical Philosophy* appeared, Joseph Gay-Lussac (1778 – 1850) published a new empirical law, the law of combining volumes. Suppose a sample of gas #1 and a sample of gas #2 combine to form a compound. Gay-Lussac measured the *volumes* of the reactants (gas #1 and gas #2) and the product *at the same pressure and temperature.* He found that the measured volumes always had the ratio of small whole numbers. For example, suppose that a sample of hydrogen and a sample of oxygen react to form water vapor, with no hydrogen or oxygen left over. The volumes of the hydrogen and oxygen samples and the volume of the water vapor produced by their reaction, all measured at the same temperature and pressure, turn out to have the ratios 2:1:2.

Three years later, in 1811, Avogadro published his conjecture that equal volumes of dilute gases at the same temperature and the same pressure contain equal numbers of particles (atoms or molecules). Dalton's axioms then imply that in the preceding example two atoms of hydrogen combine with one atom of oxygen to make two molecules of water. So each water molecule ends up with half an oxygen atom – or, as we now say, half an oxygen molecule.

But Dalton and many of his contemporaries refused to consider the possibility that the "atoms" of oxygen (or hydrogen or nitrogen or chlorine) might be what we now call diatomic molecules. Before 1925, when quantum mechanics began to explain chemical bonds and chemical reactions, chemists had no firm basis for understanding what holds molecules together. They thought in terms of inherent affinities between atoms. Such affinities, many believed, couldn't exist between atoms of the same kind.   Finally, in 1861, Stanislao Cannizzaro (1826 – 1910), ignoring this prejudice, used Avogadro's hypothesis to infer from experimental data on combining volumes a great many mutually consistent molecular formulas of gaseous elements and compounds – enough to convince most of his colleagues that the hypothesis was sound.

This brief account of the atomic hypothesis before it became the atomic fact suggests what is missing, or at least underemphasized in the conventional account of "the scientific method." Radical and consequential scientific hypotheses are rarely if ever "suggested" by experimental and observational data. They are, as Einstein put it, free creations of the human mind. And, like other creative acts, they help bring into being novel and unforeseen kinds of order.

Maxwell's Statistical Interpretation of Thermal Equilibrium

A box one centimeter on a side filled with air at standard temperature and pressure contains roughly 3×10^{19} (30 billion billion) molecules. A completely detailed Newtonian description of this collection of particles would specify, among other things, the three position coordinates and three velocity components of each molecule. Yet experiments show that two undisturbed samples of the same gas, or of the same sample at different times, have the same measurable properties if the samples have the same temperature and the same pressure. What accounts for the vast disparity between how much data we need to characterize the state of an undisturbed gas sample at the molecular level of description and how much we need to characterize the sample's macroscopic, or thermodynamic, state?

Thermodynamics interprets equilibrium states of an isolated system as states of maximum entropy. So to understand thermal equilibrium at the molecular level we need to know the molecular-level counterpart of entropy.

Daniel Bernoulli postulated that the distribution of molecular positions and velocities in a gas sample in thermal equilibrium is uniform (the same at all positions) and isotropic (the same in all directions). Now, statistical uniformity and isotropy are aspects of randomness, or lack of order, at the molecular level. This in turn may suggest that *randomness, appropriately defined, is the molecular counterpart of thermodynamic entropy and that thermal equilibrium is the condition in which the positions and velocities of gas molecules are distributed as randomly as possible.* To pursue this suggestion we need to make the notion of a distribution (of molecular positions or velocities) more precise, and we need to define randomness.

Collisions between the molecules of an enclosed, undisturbed gas sample redistribute their positions and velocities. Experiments show that an undisturbed gas sample settles into a state in which its mass density, and hence the number of molecules per unit volume, has the same value everywhere, up to measurement error. We can plausibly assume that in thermal equilibrium the gas molecules in any small region are traveling in random directions. But what about their speeds?

We might guess that in thermal equilibrium the molecules all have the same speed. But this can't be right, because when we apply Newton's laws to collisions between, say, billiard balls, we find (and experience confirms) that their post-collision speeds differ in general from their pre-collision speeds. Newton's laws require only that the combined momentum and the combined kinetic energy of colliding particles have the same values before and after a collision – provided the collision converts a negligible part of the pair's initial kinetic energy into the particles' internal energies. So molecular collisions tend to randomize the distribution of molecular speeds. Or so it may seem intuitively. But what, precisely, does *randomize* mean in this context? And what is the "most random" distribution of the velocities of molecules in an undisturbed gas sample? In a short paper that appeared in 1860 James Clerk Maxwell offered answers to these questions.

To frame the question he began by representing a gas particle's velocity by a point in what he called *velocity space*. By introducing a Cartesian reference frame – three mutually perpendicular lines passing through the same point (the coordinate origin) – we can assign a particle three Cartesian coordinates, x, y, z, and we can assign its velocity three Cartesian components: u parallel to the x axis, v parallel to the y axis, and w parallel to the z axis. We can think of u, v, w as the *position* coordinates of a particle's representative point in *velocity* space.

The velocities of particles in a macroscopic gas sample are represented by a dense swarm of points in velocity space. To represent the distribution of these points we begin by dividing velocity space into identical cells whose edges are parallel to the coordinate axes. By the "occupation number" of a cell I mean the number of particles whose representative points lie in that cell at a given moment. By the "fractional occupation number" I mean the occupation number divided by the number of particles in the sample. Maxwell assumed that there is a range of cell sizes for which the fractional occupation numbers of contiguous cells are nearly equal and each fractional occupation number is proportional to the cell volume. If these conditions are met we can approximate the fractional occupation number of a cell in velocity space with dimensions du, dv, dw centered on the point (u, v, w) by the product $f(u,v,w)$dudvdw of a smoothly varying function of position in velocity space, $f(u,v,w)$, and the cell's volume, du dv dw. We can think of the fractional-occupation-number density $f(u,v,w)$ as the mass density of a smoothly varying "probability fluid." (In the present context *probability* is synonymous with *fractional occupation* number.) The "mass" of probability fluid doesn't change with time – probability fluid is conserved – because the quantities $f(u,v,w)$ dudvdw summed over all the cells in velocity space, add up to 1.

Maxwell expressed the assumption that the particle velocities in a gas sample in thermal equilibrium are randomly distributed by two mathematical conditions:

1. *The distribution is isotropic.* This condition implies that

the value of f at a point (u, v, w) depends only the point's distance from the origin $(0, 0, 0)$ and hence only on the quantity $u^2 + v^2 + w^2$, the square of that distance. That is,

$$f(u,v,w) = F(u^2 + v^2 + w^2) \text{ for some function } F.$$

2. *A molecule's velocity components in mutually perpendicular directions are uncorrelated.* Let $g(u)du$ denote the proportion, or fraction, of the particles in the sample for which u lies in a given interval du. Then the proportions with v and w in given intervals must be $g(v)dv$ and $g(w)dw$, since there are no preferred directions in velocity space. And if velocity components in mutually perpendicular directions are uncorrelated, the proportion of the molecules in the gas sample for which u, v, w lie in given intervals must be the product of the three individual proportions:

$$F(u^2 + v^2 + w^2) = g(u)\, g(v)\, g(w).$$

This equation has the solution

$$F(u^2 + v^2 + w^2) = C^3 \exp[-\, a(u^2 + v^2 + w^2)].$$

Here a and C are constants and exp is the exponential function: $\exp(x) = e^x$, the number e raised to the power x. ($e = 2.718\ \ldots$, the limit of $(1 + 1/n)^n$ as n increases without limit.)

The fractional occupation numbers $f(u,v,w)dudvdw$, summed over all the cells in velocity space, must add up to 1. This requirement determines the constant C:

$$f(u,v,w) = (a/\pi)^{3/2} \exp[-\, a(u^2 + v^2 + w^2)] \qquad (1)$$

Maxwell's two randomness conditions therefore determine the distribution of particle velocities up to a single parameter, the constant a in (1).

Now recall Bernoulli's formula for the pressure of an ideal gas,

$$P = \tfrac{2}{3}\, n\, \langle E \rangle,$$

where n is the number of particles per unit volume, $E = \tfrac{1}{2}\, m(u^2 + v^2 + w^2)$, the kinetic energy of a particle, and $\langle E \rangle$ is the average value of E. This formula becomes the equation of state of an ideal gas, $P = nT$, if we define the temperature T of an ideal gas as ⅔ $\langle E \rangle$.

Using Maxwell's formula (1) for the distribution of particle velocities, we can calculate the average value of the squared particle velocity, and hence the average kinetic energy, which we then set equal to T. This calculation gives $a = m/2T$, so Maxwell's formula becomes

$$f(u,v,w) = (m/2\pi T)^{3/2} \exp[- E/T] \qquad (2)$$

If we want to measure temperature on the Kelvin scale, we must replace T in this formula by kT, where k is a constant, called Boltzmann's constant, whose value depends on the units in which we choose to measure mass, length, and time.

Because Maxwell's formula was not suggested by observational or experimental evidence but instead followed from mathematical conditions that Maxwell devised to make precise the notion of molecular disorder, it had a mixed reception. Most supporters of the atomic hypothesis considered the assumption that the equilibrium distribution of particle velocities doesn't favor any position or direction in space plausible, but many balked at the assumption that in thermal equilibrium the components of a gas particle's velocity are statistically uncorrelated.

Nevertheless, Maxwell and other advocates of atomism used the formula to construct quantitative theories of heat conduction, molecular diffusion, and viscous dissipation of internal motions in gases. These theories made testable prediction, which matched experiment up to experimental error.

Experimental physics didn't reach a stage in which Maxwell's formula (2) could be *directly* tested for another half-century. Since then, physicists have devised many direct tests. In one of the conceptually simplest of these a molecular beam on its way to a detector passes through narrow slits in two discs spinning at slightly different rates around a common axis parallel to the direction of the beam. The pair of spinning discs acts as a speed filter, like a pair of timed traffic lights. For a given difference between the rates at which the discs are spinning, a molecule that gets through a slit in the first disc also gets through a slit in the second disc if, and only if, its speed falls within narrow limits. This arrangement sorts molecules in the beam by their speeds.

The Doppler effect supplies another way of sorting molecules in a hot gas by their speeds. Each atom or molecule in a hot gas emits light in narrow bands of frequency, called *emission lines*. An observer equipped with a spectroscope sees light whose intensity in each narrow frequency band is the sum of the intensities of the light waves emitted in this frequency band by the gas's individual atoms. Because each atom is moving relative to the observer, its light is shifted at each frequency by an amount proportional to the line-of-sight component of its velocity. Each emission line emitted by a hot gas is accordingly shifted in frequency and broadened: its frequency is shifted by an amount proportional to the average line-of-sight velocity of the atoms; it is broadened by the atoms' thermal velocities – their velocities relative to the average velocity of the gas molecules. The profile (intensity versus frequency) of an emission line has the same shape as the distribution of line-of-sight velocities in the gas. If that distribution is Maxwellian, so is the emission-line profile.

Experiments like these show that Maxwell's formula accurately characterizes the velocity distribution of gas molecules in thermal equilibrium. Thus they support the statistical assumptions on which the formula rests – assumptions that lend specificity to the notion of a "random" distribution of molecular velocities.

In 1867 Maxwell revisited the problem of the equilibrium distribution of molecular velocities. Referring to his earlier assumption that mutually perpendicular components of molecular velocity are independently distributed, he wrote: "As this assumption may appear precarious, I shall now determine the form of the [molecular-velocity distribution] function in another way." Maxwell's new argument was part of a detailed account of how encounters between gas particles redistribute energy and momentum. It rests on the assumption that *the incoming velocities of colliding molecules are statistically uncorrelated*, a state of affairs Ludwig Boltzmann later called "molecular chaos." Like the "precarious" assumption in Maxwell's first derivation of formula (2), molecular chaos posits the absence of statistical correlations – but between the velocities of different molecules in a gas sample instead of between velocity components of the same molecule.

The 1867 paper contained another major result. Maxwell considered two intermingled samples of different gases. He assumed that initially both gases are in thermal equilibrium, each at its own temperature. He then showed that encounters between the molecules of the two gases tend to erase differences between their average kinetic energies. Earlier we saw that Bernoulli's relation $P = \frac{2}{3} n\langle E\rangle$ becomes the equation of state of an ideal gas if we define the temperature T of a gas in thermal equilibrium as a constant multiple of its average molecular kinetic energy $\langle E\rangle$. Maxwell's argument shows that the average molecular kinetic energy of a gas sample in thermal equilibrium does indeed have the defining property of temperature: energy exchanges between two intermingled gas samples with different average molecular kinetic energies tends to equalize these average molecular energies. Notice that this crucial link between Daniel Bernoulli's atomic theory of an ideal gas and the empirical gas laws was supplied not by experiments or observations but by the *predicted* outcome of a *thought* experiment.

Entropy of an Ideal-gas Sample

We've seen that certain thermodynamic properties of an ideal-gas sample in thermal equilibrium have counterparts in the atomic model:

Thermodynamic property	Atomic counterpart
Pressure	average momentum transfer per unit area
Temperature	average particle kinetic energy
Energy	total particle kinetic energy
Mass density	particle mass × average number density
Mass	total particle mass

Conspicuously absent from this list is entropy. In 1872 Ludwig Boltzmann filled this gap. He defined a statistical counterpart of thermodynamic entropy that is also a measure of randomness in systems and conditions for which entropy can't be defined. I'll refer to this measure as *randomness*. It's also called *statistical entropy*. Randomness is fundamentally a mathematical property of abstract probability distributions. It is a precise mathematical counterpart of the word's colloquial meaning.

In one way Boltzmann's definition of randomness is less general than Clausius's definition of thermodynamic entropy. Clausius's definition of entropy applies to macroscopic systems in thermal equilibrium or local thermal equilibrium. Boltzmann's definition of randomness applies to states of an ideal-gas sample characterized by arbitrary distributions of molecular position and velocity.

Boltzmann's 1872 paper also purported to demonstrate that the growth of an isolated gas sample's randomness is a consequence of Newton's laws of motion, applied to the atomic model of an ideal gas sample. Although, as we'll see, Boltzmann's proof of this claim was flawed, it broke new ground. It raised the possibility that Clausius's law of entropy change is a special case of a far broader and more deeply rooted generalization about processes that destroy (and, as I'll argue later, also create) order in the physical universe.

Boltzmann's Formula for the Randomness of a Gas Sample

We can derive a formula for the (thermodynamic) entropy of an ideal-gas sample in thermal equilibrium by combining Clausius's definition of entropy with the equations that relate the energy and pressure of such a sample to its temperature, the number of particles in the sample, and the sample's volume.

We saw earlier that the first law of thermodynamics and Clausius's definition of the entropy S imply the following formula for the change in S between neighboring equilibrium states of an enclosed gas sample:

$$dS = dU + P\,dV$$

Let's apply this formula to an ideal-gas sample. If we choose a temperature scale in which

$$U = NT \quad \text{and} \quad P = (N/V)\,T$$

then $\quad dS = N(dT/T + dV/V)$. $\hspace{4cm}$ (3)

Since $d[\log(x/x_0)] = dx/x$,

$$S = N[\log(T/T_0) + \log(V/V_0)] + \text{constant} \hspace{2cm} (4)$$

T_0 and V_0 in equation (4) are arbitrary constants. They need to be there because the argument of the logarithm function – the quantity x in the expression $\log x$ – must be a positive pure number; it mustn't depend on how we choose the units of distance, time, and mass. Any reference temperature T_0 and reference volume V_0 will do because, as we've seen, entropy is undefined up to an additive constant, and changing the values of T_0 and V_0 merely adds a constant to the right side of formula (4).

In Newtonian mechanics six quantities define the state of a particle – its three position coordinates x, y, z and its three velocity components u, v, w. If we know the laws that govern the forces acting on a particle and are given the values of these six quantities at a single moment, we can use Newton's law of motion to calculate their values at any later or earlier moment. Boltzmann united Maxwell's velocity space with position space to form a six-dimensional *state space*. The state of each particle in a gas sample containing N particles is represented by a point in this six-dimensional space, and a microstate of the whole sample is represented by a cloud of N such points.

To construct a statistical description of the state of the cloud Boltzmann divided the six-dimensional state space into cells of equal six-dimensional volume $dxdydzdudvdw$. A statistical description of the state of the cloud of representative points assigns each cell a fractional occupation number

$f(x,y,z,u,v,w) \, dxdydzdudvdw$ or $f(\tau)d\tau$, where the Greek letter τ (tau) stand for the six coordinates of a representative point in a molecule's state space and $d\tau$ stands for the cell volume $dxdydzdudvdw$ in this space. We assume we can make the dimensions of a cell so small and the value of N is so large that $f(\tau)$ varies smoothly from cell to cell, like the density of a continuous fluid.

If the assembly of N identical particles is a gas sample of volume V in thermal equilibrium, the particles are uniformly distributed in space, so the fraction of the total number of particles N contained in a region of volume dV is dV/V. The particles also have a Maxwell distribution in velocity. Finally, Boltzmann assumed that a particle's position isn't correlated with its velocity.

Then $f(\tau)d\tau = (m/2\pi T)^{3/2} \, (1/V) \exp[- E/T] \, d\tau$.

Take the negative logarithm of both sides of this formula, noting that $\log(x^k) = k \log x$, where k is any positive or negative real number:

$$-\log [f(\tau)d\tau] = E/T + \log(T/T_0)^{3/2} + \log(V/V_0), \qquad (5)$$

where $T_0^{3/2} = (m/2\pi)^{3/2} (dudvdw)$, $V_0 = dx\, dy\, dz$.

Only the first term on the right side of equation (5), E/T, varies from cell to cell (E is the kinetic energy of a molecule; its average value $\langle E \rangle$ is $3T/2$). Multiply equation (5) through by the fractional occupation number of a cell, $f(\tau)d\tau$, sum over all the cells, and denote the resulting average by angle brackets $\langle\ \rangle$:

$$\langle -\log[f(\tau)d\tau] \rangle = \langle E/T + \log(T/T_0)^{3/2} + \log(V/V_0) \rangle,$$

$$\langle -\log[f(\tau)d\tau] \rangle = \langle E/T \rangle + \log(T/T_0)^{3/2} + \log(V/V_0),$$

$$= 3/2 + \log(T/T_0)^{3/2} + \log(V/V_0)$$

Now recall that

$$S = N[\log(T/T_0)^{3/2} + \log(V/V_0)] + \text{constant.} \qquad (4)$$

From equations (5) and (4) we conclude that

$$S = N \langle -\log f(\tau)d\tau \rangle - \log f(\tau)d\tau. \qquad (6)$$

Like the statistical counterpart of the energy E of an ideal-gas sample consisting of N identical particles, the statistical counterpart of the sample's thermodynamic entropy S is N times an average value – the average value of the quantity – $\log f(\tau)d\tau$, the negative logarithm of the fractional occupation number. This quantity isn't the average value of a physical property of the gas particles. Instead it characterizes a *probability distribution* – the distribution of the points that represent single-particle states in Boltzmann's six-dimensional position-velocity space.

Boltzmann conjectured that formula (6) defines the atomic counterpart of entropy not just for samples of an ideal gas in thermal equilibrium but of *any* smooth distribution of single-particle states in position- velocity space. When the particles are uniformly distributed in space and have a Maxwell velocity distribution, the formula coincides with the formula for the thermodynamic entropy of an ideal gas. But formula (6) defines the randomness of a much wider class of macroscopic states of an undisturbed gas sample than Clausius's definition of thermodynamic entropy. For example, it defines the randomness of a gas sample in which all the particles have the same speed. Boltzmann now asked: With the statistical counterpart of entropy defined in this way, does Clausius's law of entropy change have a counterpart in his and Maxwell's statistical description of an ideal-gas sample?

Boltzmann's Transport Equation and His *H* Theorem

In his 1867 paper Maxwell used Newton's laws of motion to study the trajectories of colliding particles. He then used the results of that study to describe how particle collisions cause the average values of single-particle properties like kinetic energy to change with time. Boltzmann deepened and extended Maxwell's theory. He asked: How do particle collisions change the distribution of fractional occupation numbers in six-dimensional position-velocity space?

Boltzmann's answer to this question, his *transport equation*, equates the rate at which the fractional occupation number of a given cell in the position-velocity space of a particle in an ideal-gas sample changes with time to the difference between the rate at which particles enter the cell and the rate at which they leave the cell.

Boltzmann's transport equation became the starting point of a flourishing discipline, kinetic theory, which he and his successors used to construct testable statistical theories of processes such as heat conduction, viscous dissipation of relative fluid motions, and the diffusion of one gas through another.

Boltzmann also used his description of how molecular collisions alter the populations of cells in position-velocity space to prove his *H* theorem (Boltzmann denoted randomness by $-H$):

> *The randomness of an undisturbed gas sample not initially in thermal equilibrium increases until it reaches its largest possible value, which corresponds to thermal equilibrium.*

As you'd expect, Maxwell's form of the molecular-velocity distribution maximizes the distribution's randomness.

Boltzmann's proof of the H theorem is logically rigorous if one accepts its premises. Unlike the law of entropy non-decrease, it applies to any isolated gas sample, not just to samples in local thermal equilibrium. And unlike Clausius's law of entropy non-decrease, the H theorem says that the randomness of an undisturbed gas sample not in thermal equilibrium actually increases with time if it isn't already as large as it can be.

Criticism of the *H* Theorem

Many of Boltzmann's contemporaries rejected his statistical counterpart of entropy and his H theorem. Some of the critics rejected the atomic hypothesis and everything based on it. Advocates of Energetics, a school whose members considered the idea that heat is molecular motion to be unscientific, were especially energetic in their criticisms. The introduction to Part I of Boltzmann's *Lectures on Gas Theory* (1896)[4] contains a long list of "works on Energetics"; the foreword to Part II (1898) contains a supplementary and equally long list of "attacks on the theory of gases." Among the distinguished critics it cites are Ernst Mach, Pierre Duhem, Henri Poincaré, Robert Mayer, Wilhelm Ostwald – and V. Lenin.

But the criticism Boltzmann took most seriously was raised in 1876 by Josef Loschmidt. Loschmidt argued that the H theorem cannot show what it purports to show – that the randomness of an undisturbed gas sample not initially in thermal equilibrium increases with time – because it relies on a description of particle encounters based on Newton's laws of motion. These laws are time-reversible: they don't change when one replaces the time coordinate t by its negative, thereby reversing the direction in which t increases. So if we replace t by $-t$ in Boltzmann's account of how molecular encounters alter the populations of cells in position-velocity space *and also replace his description of the sample's initial state by a description of its final state*, his proof of the H theorem becomes a proof that the randomness of an undisturbed gas sample decreases with time.

Here's how Boltzmann describes Loschmidt's objection in his *Lectures on Gas Theory*.

---

4   Boltzmann, Ludwig 1895. *Lectures on Gas Theory*, translated by Stephen G. Brush, (New York, Dover Publications, 2011).

Let a gas be enclosed by absolutely smooth, elastic walls. Initially there is an unlikely but molecular-disordered state – for example, all the molecules have the same velocity. After a certain time the Maxwell velocity distribution will nearly be established. We now imagine that at time t, the direction of the velocity of each molecule is reversed, without changing its magnitude. The gas will now go through the same sequence of states backwards. We have therefore the case that a more probable [more highly random] distribution evolves into a less probable [less random] one, and the quantity H [the negative of randomness] increases as a result of collisions.[5]

We can rephrase Loschmidt's objection to Boltzmann's derivation of his H theorem as a question: How can molecular collisions, governed as they are by time-reversible laws, produce an irreversible macroscopic change in a gas sample? As Boltzmann points out immediately following the passage just quoted, the time-reversed picture differs from the picture envisaged in his proof of the H theorem in a crucial respect: The proof assumes (as did Maxwell's earlier proof that collisions between the molecules of intermingled gases tend to equalize their temperatures) that *the incoming (or initial) velocities of colliding molecules are uncorrelated. This assumption can't be true in the time-reversed picture imagined by Loschmidt.* Suppose, Boltzmann writes, that at some initial moment all the molecules have the same speed a. Because a collision leaves the combined kinetic energy of the collision partners unchanged, the collision partner of a molecule that emerges from its first collision with a velocity b, say, must have speed $\sqrt{(2a^2 - b^2)}$. So in the time-reversed picture, molecules with initial speeds b must collide only with molecules with speed $\sqrt{(2a^2 - b^2)}$. The assumption that the initial velocities of colliding particles are uncorrelated doesn't hold in the time-reversed picture.

Yet this assumption is crucial not only to Boltzmann's proof of the H theorem. It is a special case of an assumption that – as Boltzmann emphasized – underlies the entire molecular theory of ideal gases, from Bernoulli to Boltzmann. *The molecular theory of gases rests on the assumption that the states of individual molecules in a gas sample are uncorrelated.* If this assumption weren't true, at least to an excellent approximation, we couldn't characterize the state of an ideal-gas sample by the distribution in position-velocity space of the points that represent possible states of a single molecule. Since experience strongly supports the predictions of the molecular theory of gases, it

---

5   ibid. p. 58.

also supports the assumption that single-particle states in an isolated gas sample are statistically uncorrelated. Why then is the assumption true, or at least very nearly true?

Before addressing that question, we need to consider a theory that extends Boltzmann's theory of an undisturbed ideal-gas sample to any undisturbed system of particles whose motions and interactions conform to Newton's laws – Josiah Willard Gibbs's *statistical mechanics*.

Gibbs's Statistical Mechanics

The theory expounded by Gibbs in his *Elementary Principles in Statistical Mathematics, Developed with Special Reference to the Rational Foundations of Thermodynamics* (1902) rests squarely on ideas introduced by Maxwell and Boltzmann but takes a giant step beyond them. Boltzmann, like Maxwell, assumed that the particles of a gas sample are statistically independent; the probability of finding a particle in a particular microstate doesn't depend on the microstates occupied by the remaining $N - 1$ particles. He could therefore represent a macrostate of an undisturbed gas sample by a probability distribution of the microstates of a single gas particle. Gibbs dropped this assumption. His theory represents the macrostates of any N-particle system – not just samples of an ideal gas – by a probability distribution of N-particle microstates. (That is, it assigns a non-negative number to every possible N-particle microstate, and these numbers add up to 1.) Because Gibbs didn't assume that the particles of a macroscopic system have statistically independent probability distributions, this probability distribution – a function of $6N$ position coordinates and velocity components – doesn't factor into N identical functions of a single particle's six position coordinates and velocity components.

Gibbs represented the possible microstate of an undisturbed system of N interacting particles by a point in a $6N$-dimensional position-momentum, or phase, space. His account of how the system evolves relies on a reformulation of Newton's laws of motion published by William Rowan Hamilton in 1833.

Hamilton's version of Newtonian mechanics revealed a hidden symmetry between the roles of position and momentum in dynamical processes. It also underpins the most cogent formulation of the principles of quantum mechanics.

Following Boltzmann's example, Gibbs partitioned his phase space into identical cells. He represented a probability distribution of an undisturbed system's microstates by a large (in principle infinite) collection of points distributed among these cells, and he identified the probability that a microstate's representative point lies in a given cell with the cell's fractional occupation number (or its limiting value as the number of cells increases without limit).

Gibbs called the collection of N-particle microstates and the associated probability distribution an *ensemble*. The members of a Gibbs ensemble are *imaginary N*-particle systems. As he wrote, "Let us imagine a great number of independent systems, identical in nature but differing in phase, that is, with respect to their configuration and velocity."[6] Gibbs called the fractional occupation numbers of cells in the $6N$-dimensional phase space probabilities and denoted them by the letter P. But whereas the occupation numbers in Boltzmann's theory represent numbers of gas-particles in an actual gas sample, Gibbs's occupation numbers don't have a concrete interpretation: they refer to imaginary replicas of an N-particle system.

Having defined an ensemble, Gibbs asked: Which ensembles are the statistical counterparts of thermal equilibrium?

Because a macroscopic system has finite spatial extent and finite energy, the points that represent its possible microstates occupy a finite region of the $6N$-dimensional phase space. Since we can make the number of these points as large as we please – the points represent imaginary replicas after all – we can represent the swarm of points that represents them by a continuous fluid of variable density. If we stipulate that the total mass of the fluid is 1, the probability that a representative point lies in any given region of the phase space equals the mass of the fluid within the region. As the system evolves, the density of probability fluid changes smoothly at each point.

Gibbs now deduces from Hamilton's equations of motion that these changes have a remarkable property. Consider the moving point in 6N-dimensional position-momentum space that represents a particular evolving microstate. Gibbs proved that *the density of probability fluid **at that moving point** is constant in time.* He called this conclusion "the fundamental equation of statistical mechanics."

6   Gibbs, J. Willard, *Elementary Principles in Statistical Mathematics, Developed with Special Reference to the Rational Foundations of Thermodynamics,* (New Haven: Yale University Press, 1902)

To visualize this conclusion, imagine a swirling fluid of variable mass density in physical space. Suppose the color of the fluid varies from white to black through shades of grey in such a way that the fluid is darker where it is denser. If the motion of the fluid preserved the density of every sufficiently small fluid element, its color would remain unchanged as it moved and changed shape.

The actual motion of a fluid of variable density, such as air, doesn't have this property. For example, a sound wave in air produces alternate compressions and rarefactions; fluid elements become smaller and denser as they contract, larger and less dense as they expand. Why does the motion of probability fluid in $6N$-dimensional position-momentum space differ in this way from the motion of a real fluid like water or air in physical space?

In important ways the two kinds of motion are similar. Newton's laws of motion govern the flow of a real fluid; they also govern – though less directly, of course – the flow of probability fluid in $6N$-dimensional phase space. Moreover, both flows are conservative: one conserves the mass of physical fluid; the other conserves the mass of probability fluid. Why, then, are the volume and density of an element of probability fluid in $6N$-dimensional position-momentum space constant in time, while the volume and density of an element of a compressible fluid in physical space vary with time?

Consider an imaginary closed surface, such as a sphere, in physical space. Because mass is conserved, during any given time interval the mass of fluid inside the surface changes by an amount equal to the inflow of mass across the bounding surface (counting outflows as negative inflows). The same is true of conserved fluids in Euclidean spaces with any number of dimensions. (A Euclidean space of n dimensions is a space in which the squared distance between two points is equal to the sum of the squares of the n Cartesian-coordinate differences.) But conservation of mass does *not* entail that the volume of a fluid element (and hence its density) never changes.

Now consider a closed surface in a 6N-dimensional position-momentum space. As before, during any given interval of time the mass of probability fluid inside the surface changes by an amount equal to the inflow of mass across the surface. But the mass of probability fluid that flows across a surface element of unit area in unit time has two parts. The first part results from the fact that the 3N position coordinates of a point in 6N-dimensional position-momentum space that represents an evolving microstate change by small amounts during a short time interval. The second part is new: it results from the fact that the point's 3N momentum coordinates also change by small amounts during that time interval. When we calculate the probability current we need to add these two contributions. And when we do that, using Hamilton's version of Newton's laws, we find that the volume, as well as the mass, of an element of probability fluid centered on a point that represents an evolving microstate never changes.[7]

Earlier we denoted by τ the set of six coordinates of a point in the six-dimensional position-velocity space of a gas particle's state space. We denoted by $d\tau$ a volume element in that space; and by $f(\tau)d\tau$ the fractional occupation number, or probability, of a cell of volume $d\tau$ centered on the point with coordinates τ. Let's now use the same notation for points, volume elements, and fractional occupation numbers in the 6N-dimensional position-momentum space of an undisturbed macroscopic system containing N particles. Gibbs's "fundamental equation of statistical mechanics" then says that $f(\tau)$, evaluated at a moving point $\tau(t)$ that represents an evolving microstate of the system never changes. As before, define the randomness S of the probability distribution $\{f(\tau)d\tau\}$ as the probability-weighted average, or mean, of the negative logarithm of the probability

$$f(\tau)d\tau = \langle -\log[f(\tau)d\tau).$$

$$S = \langle -\log \langle f(\tau)d\tau).$$

Gibbs's fundamental theorem implies that if $\tau(t)$ denotes the 6N coordinates of a moving point that represents an evolving microstate

7   ibid.

of an undisturbed system, then the mass of probability fluid in an evolving cell, $f[\tau(t)]d\tau(t)$, doesn't depend on the time t.

So neither do its logarithm and the negative mean of its logarithm, the randomness S:

> The randomness of the probability distribution that characterizes an evolving macrostate of an undisturbed macroscopic system is constant in time.

In marked contrast, the entropy of an undisturbed macroscopic system may increase with time.

Boltzmann showed that an undisturbed gas sample is in thermal equilibrium when the randomness of the probability distribution of single-particle microstates has the largest value compatible with a given value of the mean particle energy $\langle E \rangle$. He also showed that the single-particle randomness, multiplied by the number of particles N in a gas sample, plays the same role in the statistical theory of an ideal-gas sample as Clausius's entropy does in thermodynamics. Gibbs proved the analogous propositions for an undisturbed system composed of particles whose motions and interactions are governed by Newton's laws of motion:

> An undisturbed macroscopic system is in thermal equilibrium when the randomness of the probability distribution of its microstates has the largest value compatible with the mean energy E of the imaginary replicas that make up the ensemble that represents the system's macrostate. Call this largest value S_{max}. S_{max} plays the same role in Gibbs's statistical description of equilibrium states as Clausius's entropy does in thermodynamics.

Gibbs also showed, as Maxwell had done for an ideal gas, that the parameter T in Maxwell's formula for the randomness-maximizing probability distribution of microstates has the defining property of temperature: when two initially undisturbed macroscopic systems characterized by randomness-maximizing probability distributions with different values of T are brought together and allowed to interact, the system with the larger value of T loses energy to the system with the smaller value, and the systems' combined randomness increases during this process.

After 1925, quantum mechanics replaced Newtonian mechanics as the fundamental theory of matter and radiation on small scales. But thanks to deep structural similarities between quantum mechanics and Hamilton's formulation of Newton's laws, the mathematical framework of Gibbs's theory survived this replacement virtually unchanged. Gibbs's theory, modified to fit the new microphysics, has succeeded brilliantly in predicting the observable properties of matter and radiation in thermal equilibrium.

It also has an important new feature. The six-dimensional position-momentum space, or phase space, of a particle governed by quantum mechanics differs from the phase space of a classical particle. We can partition a classical particle's phase space into arbitrarily small cells of equal (six-dimensional) volume. In contrast, cells in the phase space of a quantum particle have a minimum volume, determined by Werner Heisenberg's indeterminacy relations: the product of the indeterminacies of a position coordinate and the corresponding momentum component can't be less than $h/4\pi$, where h is Planck's constant. So the volume of a cell in the phase space of a quantum particle can't be less than $(h/4\pi)^3$, and the volume $d\tau$ of a cell in the $6N$-dimensional phase space of a collection of N identical quantum particles can't be less than $(h/4\pi)^{3N}$. Now recall that Boltzmann's randomness, like Clausius's entropy, is defined only up to an additive constant; only changes in randomness and entropy are well defined. In Boltzmann's theory this is because we can partition the position-momentum space of a classical particle into cells of arbitrarily small volume. The smaller we choose a cell's volume, the smaller the cells' fractional occupation numbers; the smaller the cells' fractional occupation numbers, the larger the average value of the logarithm of their reciprocals – Boltzmann's randomness. For quantum particles Heisenberg's indeterminacy relations imposes a lower limit on the volume of a cell in position-momentum space, and hence an upper limit on the randomness of an N-particle gas. *Quantum mechanics makes randomness – not just its changes – well defined.* I'll argue later that randomness is an objective property of the physical universe.

Boltzmann's molecular theory of gases replaces thermodynamics' description of the equilibrium states of ideal-gas samples by a statistical theory based on the assumption that gases consist of particles moving and interacting in ways governed by Newton's laws of motion. Gibbs's theory extends Boltzmann's theory to arbitrary macroscopic systems. Both theories have enjoyed great predictive success. Yet, as the preceding discussion has shown, they don't fit smoothly together. Nor does either theory fit smoothly with thermodynamics:

— In Boltzmann's theory the randomness of the probability distribution that characterizes a macrostate of an undisturbed gas sample containing N particles equals the randomness of the probability distribution that characterizes a macrostate of a single gas particle, multiplied by N. This is a consequence of the assumption that the motions of gas particles are statistically uncorrelated. But as discussed below, Gibbs's theory shows that this relation almost never holds. And even if it were to hold at one moment, it would immediately break down, because, as Boltzmann himself emphasized, encounters between gas particles create statistical correlations between the probability distributions of their velocities.

— How can we reconcile Boltzmann's proof that the randomness of an undisturbed gas sample not initially in thermal equilibrium always increases with the fact, pointed out by Loschmidt, that the proof rests on a *time-reversible* description of the motions of gas particles in the sample?

— How can we reconcile Gibbs's proof that the randomness of a closed system is constant in time with Boltzmann's proof that the randomness of an undisturbed sample of an ideal gas increases with time (unless it already has its largest possible value)?

— Gibbs's proof that the randomness of a closed system is constant in time also clashes with Clausius's law of entropy non-decrease. Yet, as we saw, the randomness of the probability distribution that characterizes an equilibrium state of an ideal-gas sample coincides with the system's entropy.

— Finally, Boltzmann's theory supplies a physical interpretation of the probabilities it assigns particle microstates. It locates the N points that represent the microstates of a gas sample's N particles in a *single particle's* six-dimensional position-velocity space, and it identifies the probabilities of single-particle states with the fractional occupation numbers of (sufficiently small) cells in that space. Gibbs's statistical mechanics, in contrast, leaves the probabilities it assigns the microstates of a macroscopic system uninterpreted. It identifies system microstates with fractional occupation numbers of cells in the $6N$-dimensional position-momentum space of an N-particle system, but the systems whose representative points are distributed among these cells are imaginary replicas, or, as Erwin Schrödinger called them, "mental copies" of a single, real system. "Now what on earth could it mean, physically," Schrödinger asked, "to distribute a given amount of energy over [a collection of] mental copies?"[8]

We can gain insight into these issues by using some mathematical properties of randomness stated and proved by Claude Shannon in 1948 in "*A Mathematical Theory of Communication*," the paper that launched information theory.[9]

Conditional Randomness and Correlation Information

Let $\{p_i\}$ denote a discrete set of non-negative numbers p_i that add up to 1. These are the defining properties of probabilities p_i and a probability distribution $\{p_i\}$. The index i labels an individual "event" in a discrete set of events $\{i\}$. (It can also label a possible value A_i of a *random variable* A, a mathematical object that has a discrete set of possible values A_1, A_2, ... with corresponding probabilities p_1, p_2, ... These might be the possible outcomes of a measurement of a physical quantity A.)

We interpret the quantity

---

8   Schrödinger, Erwin, *Statistical Thermodynamics*, (Cambridge University Press, 1948) p. 3.
9   Shannon, C.E., *The Bell System Technical Journal* (Vol. 27, July, October, 1948) pp. 379–423, 623–656.

$$S = -\Sigma_i \, p_i \log p_i$$

as a measure of the probability distribution's randomness.

Denote by S_{max} the largest value of S for which the set of probabilities $\{p_i$ satisfies a given constraint. For example, we might require the mean of a molecule's kinetic energy, $\Sigma_i \, p_i \, E_i$, to have a given value. Boltzmann's theory characterizes the state of thermal equilibrium of a gas sample by the probability distribution that maximizes S subject to this constraint.

We interpret the difference

$$I = S_{max} - S$$

as a measure of a probability distribution's *information*. A probability distribution is maximally informative if $S = 0$, that is, if one of the system's possible state has probability 1 and the rest have probability zero. It is least informative if S has the largest possible value that is consistent with given constraints on the probability distribution. If the number of possible states is finite and equal to n and there are no constraints on the probability distribution, the most random distribution assigns the same probability $1/n$ to each state, and $S = \log n$.

Suppose that the error δ of a particular measurement has a certain probability distribution. Assume that the mean error is 0 and that the average of the squared error, or mean square error, is some real number σ^2. The most random probability distribution of measurement errors δ – the distribution whose randomness is as large as possible subject to these conditions – is then $f(\delta) = (1/\sqrt(2\pi \, \sigma^2))$ $\exp[-\delta^2/2\sigma^2]$. The probability that the error lies between δ and $\delta + d\delta$ is $f(\delta)d\delta$. Maxwell's theory predicts that each component of a molecule's velocity in a gas sample in thermal equilibrium is distributed according to this formula. So the present definition of randomness coincides with Maxwell's.

Now consider a pair of random variables, A, whose possible values are A_1, A_2, ..., and B, whose possible values are B_1, B_2, ... We assign each pair of possible values A_i, B_j *a joint probability $P(A_i, B_j)$*. Summed over all pairs of indices i, j, these joint probabilities add up to 1. Summed over the possible values B_j of B, the joint probabilities

$P(A_i, B_j)$ add up to $P(A_i)$, the probability that A takes the value A_i regardless of the value taken by B:

$$P(A_i) = \Sigma_j\, P(A_i, B_j). \tag{a}$$

Similarly, $\quad P(B_j) = \Sigma_i\, P(A_i, B_j). \tag{b}$

We can think of the joint probabilities as entries in a table whose rows are labeled by the index i and whose columns are labeled by the index j. The entries in the ith row sum to $P(A_i)$. So if we divide each of these entries by $P(A_i)$ we get a new set of nonnegative numbers that sum to 1. These are the conditional probabilities $P(B_j|A_i)$; $P(B_j|A_i)$ is the probability that B takes the value B_j given that A has the value A_i :

$$P(B_j|A_i) = P(A_i, B_j)/P(A_i), \quad \Sigma_j\, P(B_j|A_i) = 1 \ (i = 1, 2, ...)$$

Similarly, the entries in the jth column sum to $P(B_j)$, so if we divide the ith entry in the jth column by $P(B_j)$ we get the conditional probability $P(A_i|B_j)$, the probability that A takes the value A_i given that B has the value B_j :

$$P(A_i|B_j) = P(A_i, B_j)/P(B_j), \quad \Sigma_i\, P(A_i|B_j) = 1 \ (j = 1, 2, ...)$$

Since the logarithm of a product is the sum of the logarithms of the factors,

$$\log P(A_i, B_j) = \log P(A_i) + \log P(B_j|A_i)$$
$$= \log P(B_j) + \log P(A_i|B_j)$$

If you insert these expressions for $\log P(A_i, B_j)$ into the definition of S you will find, after a short calculation:

$$S(\{P(A_i, B_j)\}) = S(\{P(A_i)\}) + \Sigma_i\, P(A_i)\, S(\{P(B_j|A_i)\})$$
$$= S(\{P(B_j)\}) + \Sigma_i\, P(B_j)\, S(\{P(A_i \mid B_j)\})$$

or, more succinctly,

$$S(A, B) = S(A) + \Sigma_i\, P(A_i)S(B|A_i)$$
$$= S(B) + \Sigma_j\, P(B_j)S(A \mid B_j)$$

The joint randomness of a pair of random variables A, B equals the randomness of one of them plus the weighted average value of the conditional randomnesses of the other.

The random variables A, B are said to be (*statistically*) *independent* or *uncorrelated* if, for every pair of indices i, j the joint probability $P(A_i, B_j)$ is the product of the individual probabilities:

$P(A_i, B_j) = P(Ai)\, P(Bj)$, just in case A, B are independent, or uncorrelated, random variables (RVs).

And this condition implies that the randomness of the joint probability distribution of uncorrelated random variables equals the sum of the randomnesses of the individual probability distributions:

$$S(\{P(Ai, Bj)\}) = S(\{P(Ai)\}) + S(\{P(Bj)\}),$$

or more succinctly,

$$S(A, B) = S(A) + S(B) \quad \text{if } A, B \text{ are uncorrelated RVs.}$$

Now consider the set of all joint probability distributions $\{P(A_i, B_j)\}$ that have a given pair of individual probability distributions $\{P(A_i)\}$, $\{P(B_j)\}$ that satisfy formulas (a), (b) above. Using a standard algorithm of elementary calculus (the method of Lagrange multipliers), one can find the joint probability distributions $\{P(A_i, B_j)\}$ that maximizes the randomness $S(\{P(A_i, B_j)\})$, or $S(A, B)$, subject to these conditions. It turns out to be the joint probability distribution $\{P(A_i)\, P(B_j)\}$.

> The randomness of the joint probability distribution of two random variables has its largest possible value if the variables are uncorrelated.

According to the preceding definition of information, the difference

$$S(\{P(A_i)P(B_j)\}) - S(\{P(A_i, B_j)\})$$

is positive unless A and B are uncorrelated, in which case it vanishes. The difference represents *correlation information* – information associated with correlations between the random variables A, B. To repeat: Unless the random variables A and B are statistically independent, the randomness of their joint probability distribution falls short of its largest possible value, the sum of the statistical entropies of the probability distributions of the individual variables, by a quantity that represents information associated with statistical correlations between the variables. (Two random variables are correlated if the probability that one of them takes one of its possible values depends on the value taken by the other random variable.)

If we have three random variables, A, B, C, a straightforward generalization of the preceding argument shows that the randomness of their joint probability distribution falls short of its largest possible value, the sum of the randomnesses of the probability distributions of the individual variables, by a quantity that represents information associated with statistical correlations between the random variables – correlations between the pairs A and B, B and C, A and C, as well as triple correlations.   With the concept of correlation information in hand, we can now resolve the conflicts between Boltzmann's and Gibbs's theories. We can also get a deeper understanding of Loschmidt's objection to the H theorem and Boltzmann's reply to that objection.

Boltzmann and Gibbs

Gibbs represented the possible macrostates of an undisturbed system of N interacting particles by probability distributions of N-particle microstates. The preceding discussion shows that (a) the randomness S of such a probability distribution takes its largest possible value when the microstates of the N particles have mutually independent probability distributions; (b) the N-particle probabilities are then products of single-particle probabilities, and S is N times the randomness of the single-particle probability distribution; and (c) the difference between S and its largest possible value represents information associated with correlations between variables that refer to different particles.

Boltzmann assumed, in effect, that correlation information is *initially* absent in an N-particle gas sample; the N-particle randomness is then initially equal to N times the single-particle randomness. Gibbs proved, however, that the N-particle randomness of any isolated system is constant in time. Boltzmann's response to Loschmidt's criticism of the H theorem shows that he understood that particle collisions create statistical correlations between particles. His proof of the H theorem shows that particle collisions *initially create correlation information* – that in an undisturbed gas sample, an increase in single-particle randomness is accompanied by an equal increase in correlation information. *Collisions transform single-particle information into correlation information.*

Boltzmann's proof of the *H* theorem would be valid if, and only if, the correlation information created by the decay of single-particle information immediately left the system – if, say, it leaked into the system's surroundings.

And if correlation information tended to leak into a nominally undisturbed system's surroundings, the following discussion shows that we would also be able to resolve the conflict between Clausius's law of entropy non-decrease and Gibbs's theorem that the randomness of the probability distribution that characterizes a macroscopic system's macrostate is constant in time. We'd then be led to ask: *Under what conditions does correlation information leak out of a nominally undisturbed macroscopic system? And is there any reason to suppose that such conditions regularly occur?*

Gibbs's Theorem and Clausius's Law

Gibbs's definition of the randomness of an undisturbed macroscopic system applies to a wider class of macroscopic systems than Clausius's definition of thermodynamic entropy. Yet Gibbs proved that the randomness of an isolated, or undisturbed, system is constant in time, whether or not the system is in a state of thermal equilibrium. In contrast, the system's thermodynamic entropy can increase with time if the system is not initially in thermal equilibrium. Gibbs's proof that the randomness of an undisturbed system is constant in time is simple and straightforward. His "fundamental equation of statistical mechanics" shows that both the volume and the "mass" of an element of probability fluid centered on a point that represents an undisturbed system's evolving microstate are constant in time.

In Chapter XII of his *Elementary Principles of Statistical Mechanics* Gibbs addresses this conflict between his theory and Clausius's law of entropy non-decrease. He argues that the probability fluid in 6*N*-dimensional phase space changes with time like a stirred liquid. Suppose the liquid contains coloring matter of variable density.

> If we give the liquid any motion whatever ... the density of the coloring matter at any same point [i.e., any point moving with liquid]

will be unchanged.... Yet no fact is more familiar to us than that stirring tends to bring a liquid to a state of ... uniform densities of its components. ... [10]

Imagine the region occupied by the liquid to be divided into cells of equal volume. No matter how small we make these cells, the distribution of coloring matter will eventually appear to be uniform on that scale. In language introduced by Gibbs's successors, the "coarse-grained" probability distribution becomes increasingly uniform while the "fine-grained" distribution never changes. Some of Gibbs's successors therefore proposed to identify thermodynamic entropy with the coarse-grained randomness of the probability distribution that characterizes a macrostate of an undisturbed system in Gibbs's theory..

Gibbs's argument doesn't, however, provide a criterion for the scale of coarse-graining. More important, while his analogy between the flow of probability fluid in phase space and the flow of a colored liquid is illuminating, it doesn't supply a *reason* for the temporal asymmetry of the process. Because the equations that govern the motion of an ideal fluid are time-reversible, any description of the mixing process is likewise reversible if we exclude the effects of molecular diffusion (as Gibbs does). In a single brief passage Gibbs addresses this problem. He suggests that the conflict we've been discussing reflects an inability of mathematical models to capture physical reality:

> But while the distinction of prior and subsequent events may be immaterial with respect to mathematical fictions, it is quite otherwise with respect to the events of the real world. It should not be forgotten, when our ensembles are chosen to illustrate the probabilities of events in the real world, that while the probabilities of subsequent events may often be determined from the probabilities of prior events, it is rarely the case that the probabilities of prior events can be determined from those of subsequent events, for we are rarely justified in excluding the consideration of the antecedent probability of the prior events. [11]

Whereas a realistic picture of the physical world distinguishes between the directions of the past and the future, Gibbs's statistical mechanics, like Boltzmann's kinetic theory of gases, rests on mathematical laws – Newton's laws of motion – that don't make this distinction.

---

10   Gibbs *Elementary Principles* pp. 144-5.
11   Gibbs *Elementary Principles* p. 150.

Toward a Resolution

Although Clausius's law of entropy non-decrease rests on an empirical generalization (that no cyclic engine can convert heat drawn from a single source into mechanical energy and leave the rest of the world unchanged), no physical law has been confirmed more strongly or in more different ways. Yet Gibbs's statistical mechanics, which supplies what Gibbs in the subtitle of his book called "a rational foundation" for thermodynamics, predicts that the statistical counterpart of the entropy of an undisturbed system is constant in time. If both Clausius and Gibbs are right, as I think they are, it follows that the systems for which Clausius's law holds aren't, strictly speaking, undisturbed. For if they were, and if the domain of statistical mechanics includes the domain of thermodynamics (as it surely does), Gibbs's theorem implies that the law of entropy non-decrease is false. So we're led to ask: When are we justified in treating macroscopic systems as if they didn't interact at all with the outside world?

We can think of any macroscopic system as part of a larger macroscopic system. If we take *that* system to be large enough, we are surely justified in neglecting *its* interaction with its environment when we describe what is happening in the original system. Now recall Gibbs's remark that "while the distinction of prior and subsequent events may be immaterial with respect to mathematical fictions, it is quite otherwise with respect to the events of the real world." Gibbs's statistical mechanics applies to *undisturbed* macroscopic systems – systems that do not interact at all with the outside world. The Second Law, in contrast, refers *directly* to the world of experience.[12] It refers neither explicitly nor implicitly to undisturbed systems, which are mathematical fictions. This line of reasoning suggests that by investigating the environments in which macroscopic systems are usually embedded and how the nominally isolated systems of experimental macrophysics interact with these environments, we may be able to do three things: reconcile Gibbs's theorem on the constancy of entropy's statistical counterpart with the law of entropy non-decrease; reconcile Gibbs's theorem with

12   "We, therefore, put forward the following proposition as being given directly by experience: It is impossible to construct an engine which will work in a complete cycle, and produce no effect except the raising of a weight and the cooling of a heat-reservoir." Max Planck, Treatise on Thermodynamics, trans. Alexander Ogg. Third edition translated from the seventh German edition. Dover Publications, New York, 1945.

Boltzmann's H theorem and with the postulate of molecular chaos, which underpins both Maxwell's and Boltzmann's gas theories; and counter Loschmidt's objection to the H theorem.

Physicists recognized early on that they could gain insight into the mathematical regularities that lie behind appearance by minimizing interactions between the objects they studied in their laboratories and the outside world, and by ignoring these interactions in their theories. Galileo's experimental and theoretical studies of the (almost) frictionless motion of bodies sliding down inclined planes illustrate this strategy. But whereas macroscopic systems can be effectively insulated against exchanges of heat, mechanical energy, and momentum with their environments, it's much harder to insulate them against exchanges of *information* (in the technical sense defined earlier) with their environments. Suppose now that nominally undisturbed macroscopic systems are normally embedded in highly random environments, and consider an isolated system composed of a nominally undisturbed macroscopic system S and an extensive (but bounded) random environment 00 – a system characterized by a probability distribution of microstates whose randomness has its largest possible value. By Gibbs's theorem, the randomness of the combined system $S + E$ is constant in time. But, as Boltzmann's proof of his H theorem implies, collisions between molecules of S and molecules of E at the interface between S and E create molecular correlations. Information associated with these correlations flows from S to E. Hence the randomness of S increases, while the randomness of E decreases by an equal amount. The transfer of correlation information from S to its environment preserves the mutual statistical independence of the single-particle probability distributions and causes the randomness of the N-particle distribution to increase steadily with time – a stronger – as well as more widely applicable –statement than that provided by the law of entropy non-decrease.

But what justifies the assumption that (actual) macroscopic systems are embedded in random environments? The assumption of primordial randomness implies that

> A complete description of a macroscopic system's initial
> state contains just the information created by the system's
> history, which includes an account of how the initial state
> was prepared.

This supplies a framework – but only a framework – for address-ing what the physicist Nico Van Kampen[13] has called "the main problem in the statistical mechanics of irreversible processes:" What determines the choice of macrostates and macroscopic variables?

Van Kampen has also emphasized the role of "repeated random-ness assumptions" in theories of stochastic processes:

> This *repeated randomness assumption* is drastic but indispensable whenever one tries to make a connection between the micro-scopic world and the macroscopic or mesoscopic levels. It appears under the aliases "*Stosszahlansatz*," "molecular chaos," or "random phase approximation," and it is responsible for the appearance of irreversibility. Many attempts have been made to eliminate this assumption, usually amounting to hiding it under the rug of mathematical formalism.[14]

To derive his transport equation and his *H* theorem, Boltzmann had to assume (following Maxwell) that the initial velocities of colliding molecules in a closed gas sample are uncorrelated. This assumption can't hold for non-equilibrium states, because when the information that characterizes a non-equilibrium state decays, resid-ual information associated with molecular correlations comes into being at the same rate. Indeed Poincaré proved that if you wait long enough, the (classical) microstate of an undisturbed system whose particles move and interact in ways governed by Newton's laws of motion eventually returns to a state that approximates the system's initial state arbitrarily closely. However, because a gas sample's res-ervoir of microscopic information is huge, molecular chaos may prevail to a good approximation for periods much shorter than the Poincaré recurrence time, which is typically much greater than the age of the universe.

Another approach to the problem of justifying repeated random-ness assumptions starts from the remark that no gas sample is an island unto itself. Almost sixty years ago, J. M. Blatt[15] argued that the fact that actual gas samples interact with their surroundings justifies

13   N. G. van Kampen in *Fundamental Problems in Statistical Mechanics, Proceedings of the NUFFIC International Summer Course in Science at Nijenrode Castle*, The Netherlands, August, 1961, compiled by E. G. D. Cohen (Amsterdam, North-Holland, 1962), especially pp. 182-184.

14   N.G. van Kampen, *Stochastic Processes in Physics and Chemistry*, 3d edition (Amsterdam, Elsevier, 2007), p. 58.

15   J. M. Blatt, *Prog. Theor. Phys.* 22, 745 (1959).

the assumption that correlation information is permanently absent in a nominally closed gas sample. The walls that enclose a gas sample are not perfectly reflecting. When a gas molecule collides with a wall, its direction and its velocity acquire tiny random contributions. These leave the single-particle probability distribution virtually unaltered, but they alter the histories of individual molecules, thereby disrupting multi-particle correlations. Blatt distinguished between states of true equilibrium, characterized (in the vocabulary of the present essay) by an information-free probability distribution, and quasi-equilibrium states, in which single-particle information is absent but correlation (or residual) information is present. With the help of a simple thought experiment, he argued that collisions between molecules of a rarefied gas sample and the walls of its container cause an initial quasi-equilibrium state to relax into true equilibrium long before the gas has come into thermal equilibrium with the walls. Earlier, P.J. Bergmann and J.L. Lebowitz[16] constructed and investigated detailed mathematical models of the relaxation from quasi-equilibrium to true equilibrium through external "intervention." More recently, T.M. Ridderbos and L.G. Redhead[17] have expanded Blatt's case for the interventionist approach. They constructed a simple model of a famous experiment, the spin-echo experiment[18], in which a macroscopically disordered state evolves into a macroscopically ordered state. They argued that in this experiment (and more generally) interaction between a nominally isolated macroscopic system and its environment mediates the loss of correlation information.

Blatt noted that this "interventionist" approach "has not found general acceptance."

> There is a common feeling that it should not be necessary to introduce the wall of the system in so explicit a fashion. ... Furthermore, it is considered unacceptable philosophically, and somewhat "unsporting," to introduce an explicit source of randomness and stochastic behavior directly into the basic equations. Statistical mechanics is felt to be a part of mechanics, and as such one should be able to start from purely causal behavior.[19]

---

16   P. J. Bergmann and J. L. Lebowitz, *Phys. Rev.* 99, 578 (1955); J. L. Lebowitz and P. J. Bergmann, Annals of Physics 1, 1 (1959).

17   T. M. Ridderbos and M. L. G. Redhead, *Found. Phys.* 28, 1237 (1988).

18   E. L. Hahn, *Phys. Rev.* 80, 580 (1950).

19   J. M. Blatt, *Prog. Theor. Phys.* 22, 745 (1959), p. 747.

Derivations of Boltzmann's H theorem and its generalization that rely on decoherence[20] exemplify and significantly extend the interventionist approach. These derivations assume that macroscopic systems are initially in definite quantum states but interact with random environments. Under these conditions, environmental interactions transfer information very effectively from the system to its surroundings. Microscopic information that is wicked away from the system disperses outward, eventually getting lost in interstellar and intergalactic space.

Because the microstates of undisturbed systems evolve reversibly, a theory that assigns macroscopic systems (or the universe) definite quantum states cannot provide a framework for theories that distinguish in an absolute sense between the two directions of time. The historical approach sketched in this book offers such a framework because by linking the initial states of macroscopic systems to states of the early universe it supplies a cosmological context for theories of irreversible processes. Because the probability distributions that characterize the initial states of macroscopic systems depend on their histories, there can be no genuine laws about initial conditions, only historical generalizations. For example, macroscopic systems cannot usually be prepared in definite quantum states; but physicists have succeeded in preparing superconducting quantum interference devices (SQUIDs) in superpositions of macroscopically distinct quantum states. Again, macroscopic systems are usually microscopically disordered; but Hahn's spin-echo experiment showed that this is not necessarily the case. A historical narrative based on the strong cosmological principle and the assumption of primordial randomness predicts that macroscopic systems are *usually* embedded in random environments and that their initial states *usually* contain only information associated with the values of macroscopic variables. However, as the spin-echo experiment shows, macroscopic systems *can* be prepared in states that do contain microscopic information.

---

20   M. Schlosshauer, *Decoherence and the Quantum-to-Classical Transition*, corrected 2d printing (Berlin, Springer, 2008).

VI

A Unified Picture
of the
Physical World

The Measurement Problem

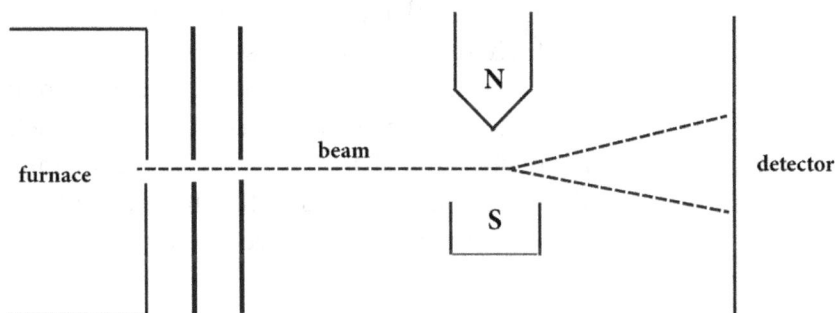

Figure 2 . The Stern-Gerlach experiment

Some atoms – silver, for example – have magnetic moments: in a magnetic field they behave as if they were tiny bar magnets. In 1922, three years before the birth of quantum mechanics and two years before Samuel Goudsmit and George Uhlenbeck proposed that electrons have an intrinsic spin and an intrinsic magnetic moment, physicists Otto Stern and Walter Gerlach devised a now classic experiment to measure the strength of a silver atom's magnetic moment. (Because silver atoms have a single valence electron, they have a magnetic moment equal to the electron's magnetic moment.) Their experimental setup included a vertical magnetic field whose field-lines diverged from top to bottom. When a silver atom traveling in a horizontal direction enters such a field, classical electromagnetic theory predicts that it is deflected upward or downward by an amount proportional to the cosine of the angle its magnetic moment makes with the field. In the Stern-Gerlach experiment (Figure 2) the atom, after emerging from the diverging magnetic field, strikes a detector, where it leaves a permanent trace. The height of this trace above or below the point at which the atom would have struck the detector if it hadn't been deflected by the magnetic field is proportional to the vertical component of the atom's magnetic moment. So if we knew the initial orientation of the atom's magnetic moment we could infer its magnitude from the height of the trace. This is the quantity Stern and Gerlach set out to measure.

The experiment was done not with a single silver atom but with a narrow beam of silver atoms. The atoms had nearly the same horizontal velocities but randomly oriented magnetic moments. So Stern and Gerlach expected that the heights of the traces produced by individual atoms would be distributed smoothly along a vertical strip, whose length would be a calculable multiple of a silver atom's magnetic moment. To their surprise the traces of individual atoms clustered around the points that corresponded to the two extreme orientations of the magnetic moment, as if half of the magnetic moments had been initially parallel to the field, half antiparallel. Yet the conditions of the experiment ensured that before the atoms entered the field their magnetic moments were randomly oriented. What was going on?

Quantum mechanics eventually answered the question. Its laws predict that under the conditions of the Stern-Gerlach experiment the magnetic moment of a silver atom must exhibit one of two equal values, parallel or antiparallel to the magnetic field. But while this prediction, along with countless other similar predictions, agrees with experiment, physicists still – after more than 90 years – don't agree about how to interpret the theory on which it rests. They disagree about how that theory relates to classical physics and the world of experience. That disagreement in turn reflects, in part, a diversity of opinion about the nature of chance and its role in physical theories – questions that also underlie the issue of free will versus determinism.

The Classical World and the Quantum World

Classical physics both refines and extends commonsense notions about the physical world. The world of classical physics is made up of objects with measurable properties, usually called "dynamical variables" or just "physical quantities." Physical quantities have definite values, which represent the outcomes of ideal error-free measurements and which define the physical state of the system they refer to. For example, at each moment a classical particle has a definite position, specified by its three position coordinates in some coordinate system, and a definite velocity, specified by its three velocity components in that coordinate system; and these six quantities define its physical state, which varies smoothly with time in a way governed by Newton's laws of motion and appropriate force laws, including Newton's law of gravitation.

Some of the physical quantities of classical physics have counterparts in quantum theory, where they go by the same names. Among these properties are position, velocity, energy, linear momentum, and angular momentum. Many relations between classical quantities also have quantum counterparts. For example, the formulas that express a particle's momentum, kinetic energy, potential energy and angular momentum in terms of its position and velocity have the same form in classical physics and quantum physics.

But the mathematical objects that represent the quantum counterparts of classical quantities don't always have definite numerical values when the system they refer to is in a definite state. For example, while the electron in a hydrogen atom can be in a state of definite energy, neither the quantum counterparts of its position coordinates nor the quantum counterparts of its momentum components have definite values in such a state.

Because quantum dynamical variables can't in general be assigned definite numerical values, the outcome of an ideal error-free measurement of a quantum dynamical variable that refers to a quantum system in a definite state isn't predictable in general. It does, though, have a predictable set of *possible values*. These are called the physical quantity's *eigenvalues*. For example, the Stern-Gerlach experiment shows that the magnetic moment of a silver atom has just two eigenvalues, $+h/4\pi$ and $-h/4\pi$, where h is the constant introduced in 1900 by Max Planck in his semi-empirical formula for the frequency distribution of thermal radiation.) Each eigenvalue of a physical quantity is associated with one or more *eigenstates*.

> An ideal measurement of a physical quantity has a predictable outcome just in case the system to which the quantity refers is in one of the quantity's eigenstates. The measurement then yields the corresponding eigenvalue.

This statement is the most basic of the rules that link the mathematical formalism of quantum mechanics to the mathematical formalisms of classical theories and to experience.

Different physical quantities that refer to the same system needn't, and in general don't, have the same eigenstates. For example, the energy eigenstates of the electron in a hydrogen atom are not eigenstates of either the electron's position coordinates or its velocity components. Even if an electron is in a state of definite energy it doesn't have a definite orbit.

An undisturbed, or isolated, system needn't be in an eigenstate of *any* of the system's physical quantities, including its energy. But every state of the system can be expressed as a *superposition* of the eigenstates of any physical quantity that pertains to the system. That is, the wavefunction (or state vector) that represents a given state can be expressed as a linear combination – a sum of multiples – of wavefunctions (or state vectors) that represent eigenstates of any given physical quantity that pertains to the system. The coefficients in such linear combinations are complex numbers, numbers of the form $a + bi$, where a and b are real numbers and i denotes the square root of -1. For example, we can express the wavefunction that represents an arbitrary state of the electron in a hydrogen atom as a sum of complex multiples of wavefunctions that represent states of definite energy.

The mathematical language of quantum theory expresses these highly non-intuitive (and philosophically opaque) properties of physical quantities and physical states of molecules, atoms, subatomic particles, and other quantum systems precisely and unambiguously. These are the language's main features:

> *Quantum states.* Every isolated, or undisturbed, physical system has its own state space. The mathematical objects

that inhabit state space are called wavefunctions or, more generally, state vectors. Each state vector represents a possible state of the system. For example, an electron has two possible spin states, represented by a pair of state vectors. If a vertical magnetic field is present, an experiment designed to exhibit an electron's magnetic moment, which is proportional to its spin, will show either that the magnetic moment and the spin point up or that they point down. Spin-up and spin-down are the only two possible outcomes of such an experiment, and in a series of identical experiments they occur equally frequently.

State vectors combine with one another and with numbers much like vectors in Euclidean space.

Vectors and their algebra. Let **u** denote a vector in Euclidean space. It is defined by its direction and length. We can picture it as a free-floating arrow, untethered to any particular point. If c is a positive real number, c**u** is also a vector, represented by an arrow whose length is c times the length of the arrow that represents **u**. If c is a negative real number, c**u** is the vector whose direction is opposite to the direction of u and whose length is $|c|$ times the length of **u**, where $|c|$ denotes the absolute value of c. If **u₁** and **u₂** are any two vectors, their sum **u₁**+ **u₂** is represented by an arrow you can construct by joining the arrows that represent **u₁** and **u₂** end to tip in either order (while preserving their lengths and directions), then joining the free end to the free tip.

State vectors in an undisturbed physical system's state space have exactly the same algebraic properties, except that the constant c in the preceding account is a *complex* number. Thus if ψ_1 and ψ_2 are state vectors in the state space of a given system and a and b now denote any complex numbers, then $a\psi_1 + b\psi_2$ is also a state vector in the same space, representing one of the system's possible states. The state vector $a\psi_1 + b\psi_2$ represents a *superposition* of the states represented by ψ_1 and ψ_2.

Superpositions of quantum states have no counterpart in the language of classical physics (or in ordinary language). They are responsible for the most counterintuitive features of the quantum world. For example, the cat in Erwin Schrödinger's famous thought experiment is in a superposition of the states "dead" and "alive" until its interaction with the world of experience puts it into one state or the other, just as in the Stern-Gerlach experiment a silver atom's interaction with the measuring apparatus puts into a definite classical state, "up" or "down."

> *Observables.* The language of quantum physics represents physical quantities by mathematical objects called *observables*, a subclass of a broader class of objects called *operators*. Like the variables that represent physical quantities in Newtonian physics, observables (as well as other operators) can be added to one another, multiplied by complex numbers, and multiplied by one another. And relations between Newtonian variables have counterpart in relations between the corresponding observables. *But multiplication of operators (including observables) is non-commutative*: if O_1 and O_2 are operators, O_1O_2 need not equal O_2O_1. In particular, the observables that represent a particle's position coordinates don't commute with the operators that represent the corresponding momentum components. Heisenberg's indeterminacy relations are a consequence of the fact that "conjugate" physical quantities, like a position coordinate and the corresponding momentum component or like a particle's energy and its time coordinate, don't commute.

Physics texts and papers in physics journals often distinguish the quantum counterparts of classical variables by adding a circumflex ^ or a suffix, like $_{op}$, to the symbol that represents the corresponding classical variable, if there is one, writing $\hat{E}$ or E_{op} for the quantum counterpart of energy E. But physicists customarily use the same words, such as position, momentum, and energy, for classical variables and their quantum counterparts, relying on context to distinguish between them. In nontechnical contexts the failure to distinguish between classical variables and their quantum counterparts can lead to confusion. Consider the Stern-Gerlach experiment. Stern and

Gerlach thought of silver atoms as classical objects. They thought they were measuring the vertical position coordinates of silver atoms in the beam by recording the positions at which the atoms struck a detector. In 1923, when Stern and Gerlach carried out their experiment, the phrases "the vertical position coordinate of a silver atom" and "the height at which an atom strikes the detector" referred to the same physical quantity, the z coordinate of a silver particle. But once quantum mechanics had been formulated as a mathematical theory, this was no longer true. "The vertical position coordinate of a silver atom" became a quantum observable, represented by an operator, while "the height at which an atom strikes the detector" remained a classical quantity. What Stern and Gerlach measured, and what the classical theory of the motion of a tiny bar magnet in an inhomogeneous magnetic field (incorrectly) predicted, were values of a classical position coordinate. What quantum theory predicts, through a rule that links the quantum world to the world of experience, is the average value of the *classical counterpart* of a quantum observable.

Before they strike the detector, each silver atom is in a superposition of two distinct position states, corresponding to the fact that a silver atom's magnetic moment has just two eigenstates. But the atom isn't in two places at the same time. Nor is it strictly accurate to say that the detector records the atom's position. Before the atom interacts with the detector its position is represented by an operator and doesn't have a definite value. Afterwards the same atom does have a definite position, which coincides with a spot on a photographic plate.

More generally, in a quantum measurement a system in a definite quantum state interacts with a macroscopic measuring device, such as a photographic plate or a particle detector, that is in a definite classical state. The interaction causes the combined system, consisting of the measured system plus the measuring device, to evolve rapidly from its initial state, in which the measured system is in a definite quantum state and the measuring device is in a definite classical state, to a final joint state, in which the measured system is in an eigenstate of the measured quantity and the measuring device is in a corresponding "pointer state." The rapid evolution of the combined system during a quantum measurement is sometimes called a "jump."

Here's how Dirac describes the measurement process in the fourth edition of *The Principles of Quantum Mechanics* (1967; first edition 1930):

> When we measure a real dynamical variable ξ, the disturbance involved in the act of measurement causes a jump in the state of the dynamical system. From physical continuity, if we make a second measurement of the same dynamical variable ξ immediately after the first, the result of the second measurement must be the same as that of the first. Thus after the first measurement has been made, there is no indeterminacy in the result of the second. Hence, after the first measurement has been made, the system is in an eigenstate of the dynamical variable ξ, the eigenvalue it belongs to being equal to the result of the first measurement. This conclusion must still hold if the second measurement is not actually made. In this way we see that *a measurement always causes the system to jump into an eigenstate of the dynamical variable that is being measured, the eigenvalue this eigenstate belongs to being equal to the result of the measurement.*[1] (italics added)

Dirac doesn't describe the rapid evolution of the combined system in a quantum measurement. He fills this gap by what he calls a "general assumption." Consider an observable O and a physical state s of the system of which O is a property. Dirac's general assumption *equates the average outcome of a large number of identical ideal measurements of O when the system O pertains to is in state s to a real number constructed according to a simple rule from the operator O and the state vector s.*

Dirac's general assumption contains – and extends in the simplest possible way – our earlier basic rule: if a system is in an eigenstate of the observable O, an ideal measurement of O yields the corresponding eigenvalue of O. It implies further that if the system isn't in an eigenstate of O, the outcome of an ideal measurement is one of O's eigenvalues. Finally, it yields a formula for the relative frequencies of the possible measurement outcomes in an infinite series of identical measurements of O when the system is in state s.

Dirac's general assumption links the mathematical formalism of quantum mechanics to measurement outcomes efficiently and unambiguously. It has enabled physicists to carry out experimental

---

1   Dirac, P.A.M., *The Principles of Quantum Mechanics*, fourth edition (revised), (Clarendon Press, Oxford, 1967), p. 46.

tests of quantum theories of molecules, atoms, and subatomic particles and of macroscopic objects such as crystals, conductors, and semi-conductors – objects whose observable properties depend critically on their quantum microstructure. Up until now all such predictions have been confirmed to the accuracy of the models and calculations on which they rest.

What is missing from Dirac's account of measurement is a description of the measurement process itself – the rapid evolution, during a measurement, of the joint state of the combined system. Can quantum theory supply such a description? This is the measurement problem. In a recent popular article, "The Trouble With Quantum Mechanics," the physicist Steven Weinberg succinctly defines the core of the problem:

> If we regard the whole process of measurement as being governed by the equations of quantum mechanics, and these equations are perfectly deterministic, *how do probabilities get into quantum mechanics?*[2]

"The Quantum-to-Classical Transition"

Introductory textbooks on quantum mechanics usually have a section entitled "The Hydrogen Atom." The hydrogen atom the textbooks describe is a fictitious object because, unlike actual hydrogen atoms, it doesn't interact with the outside world. Yet the quantum theory of an undisturbed hydrogen atom accurately predicts the measurable properties of real hydrogen atoms, such as the frequencies at which a hot sample of atomic hydrogen gas emits light and a cool sample absorbs light.

In contrast, an account of quantum measurement that neglects the interaction between the measuring device and its environment fails spectacularly. For, as Weinberg emphasized, an undisturbed quantum system, like an undisturbed classical system, evolves deterministically; but the measurement postulate, Dirac's "general assumption," predicts (and experiments confirm) that in general a quantum measurement has more than one possible outcome. Thus we can neglect the interaction between a physical system and its environment, at least in a first approximation, in some circumstances but not in others. What, precisely, are these circumstances?

An undisturbed quantum system can be in a state that doesn't change with time and in which its energy, like the energy of an undisturbed classical system, has a definite value. But unlike an undisturbed classical system, whose energy has a continuous range of possible values, an undisturbed quantum system that occupies a bounded region of physical space has a discrete set of possible energies and a corresponding discrete set of possible quantum states. In 1970 physicist Hans Dieter Zeh pointed out that *we can assign the system a definite quantum state if and only if the difference between the energy of that state and the energies of neighboring states greatly exceeds the energy associated with the system's interaction with its surroundings.* If this condition isn't fulfilled, transitions between neighboring energy levels occur so rapidly that the system can no longer be assigned a state of definite energy.

Atoms and molecules in dilute gases satisfy this condition because the intervals between their low-lying energy levels typically exceed the energies associated with the interaction between an atom or molecule and its neighbors. In contrast, as I mentioned earlier, macroscopic systems have very closely spaced energy levels. Under ordinary experimental conditions the intervals between adjacent energy levels of a macroscopic object are *much* smaller than the energy associated with the object's interaction with its environment. So even if the laws of quantum mechanics apply to isolated, or undisturbed, macroscopic objects, as experimental evidence strongly suggests they do, these objects cannot, under ordinary conditions, be assigned definite quantum states.

This remark is the starting point of *decoherence theories* of quantum measurement. These theories seek to refine and improve an account of quantum measurement proposed in 1932 by the mathematician John von Neumann. Von Neumann set out to replace Dirac's general assumption by an account of the measurement process itself. He idealized the combined system in a quantum measurement – the measured system plus the measuring apparatus – as an undisturbed system in a definite quantum state. He also assumed that if the measured system is initially in an eigenstate of the measured observable, the measurement leaves it in that state and puts the measuring apparatus into a corresponding *quantum* state that is *macroscopically* distinguishable from the quantum states associated with other possible measurement outcomes. If the measured system's initial, or pre-measurement, state is not an eigenstate of the measured observable, the rules of quantum mechanics allow us to represent it as a superposition of eigenstates of the measured observable – a sum of complex multiples of the state vectors that represent the measured observable's eigenstates. Schrödinger's equation then predicts that the combined system evolves into the same superposition of combined-system states. Each state of the combined system in this superposition is a possible outcome state, in which the measured system is in one of the eigenstates of the measured observable and the measuring apparatus is in a correlated quantum state. Moreover, the squared magnitude of each coefficient in the superposition is the probability of the corresponding measurement outcome, as predicted by Dirac's general assumption.

But Dirac's general assumption implies, and experiment confirms, that a quantum measurement leaves the measured system in *one or another* of the measured quantity's eigenstates, not in a superposition of all of them.

There's a more subtle difficulty with von Neumann's account of a quantum measurement as I've described it so far. Each coefficient in the superposition that represents the measured quantum system's initial state is a complex number; it has a magnitude and a phase.[3] A measurement governed by Dirac's general assumption preserves information about the magnitudes of the coefficients in the superposition that represents the measured quantum system's initial state: their squares are the probabilities of the outcomes represented by the state vectors they multiply.

But Dirac's account of a quantum measurement implies that it destroys information about the relative phases of the coefficients. In von Neumann's account, a quantum measurement preserves information about both the magnitudes *and* the relative phases of the coefficients in the superposition that represents the evolved joint state of the measured system and the measuring apparatus.

---

3   Consider a complex number $a + bi$, where a and b are real numbers and i is the square root of –1. We can think of a and b as the rectangular coordinates of a point in a plane, the complex-number plane. The angle between the x-axis and the line joining the origin to the point that represents a complex number is called the *phase* of that number. Two complex numbers have the same phase if and only if the line joining their representative points passes through the origin of the complex-number plane.

There's also a more subtle difficulty with von Neumann's account: It contradicts the fact, stressed by Niels Bohr, that the measuring apparatus in a quantum measurement is a macroscopic system in a definite *classical* state. Von Neumann's account of a quantum measurement assumes that the measuring apparatus is initially in a definite quantum state. Decoherence theories emend von Neumann's account not by addressing this difficulty but by postulating that the measuring apparatus interacts with a random environment (in calculations, either thermal radiation or a tenuous gas). Instead of assuming, as von Neumann did, that the combined system is undisturbed (and hence in a definite quantum state), decoherence theories assume that a supersystem consisting of the combined system plus a bounded portion of the postulated random environment can be treated as an undisturbed system in a definite quantum state. A quantum measurement then transfers information about the relative phases of the state vectors of the combined system (measured system plus measuring device) to the environment. Model calculations by Wojciech H. Zurek, Erich Joos and Hans Dieter Zeh, by Maximilian Schlosshauer, and by other authors, have shown that interaction between the measuring apparatus and the random environment rapidly decoheres macroscopically separated states of the combined system, randomizing the relative phases of the coefficients in the superposition of combined-system state vectors predicted by von Neumann's account:

> Formally, decoherence can be viewed as a dynamical filter on the space of quantum states, singling out those states that, for a given system, can be stably prepared and maintained, while effectively excluding most other states, in particular, nonclassical superposition states of the kind popularized by Schrödinger's cat.[4]

By predicting the loss of coherence between different possible outcomes of a quantum measurement, decoherence theory improves von Neumann's account. But as the authors mentioned above have stressed, it doesn't predict that quantum measurements have definite but unpredictable outcomes. And, as I've already emphasized, it ignores Bohr's requirement that the measuring apparatus is in a definite classical state before and after a quantum measurement.

---

4   Schlosshauer, Maximilian, "The quantum-to-classical transition and decoherence," arXiv:1404.2635v1 [quant.phys], 9 April 2014.

Von Neumann's Collapse Hypothesis and Attempts to Avoid It

Von Neumann completed his account of quantum measurement by postulating that no sooner does the superposition of outcome states produced by interaction between the measured system and the measuring apparatus come into being than it "collapses" randomly onto one of these outcome states, the probability of each possible outcome being equal to the squared magnitude of the corresponding coefficient in the superposition. With this additional assumption, von Neumann's account reproduces – but doesn't explain – Dirac's general assumption. It replaces Dirac's "jump" of the joint state of the combined system by a "collapse" of the combined system's wavefunction. Von Neumann assumed that this collapse is a real physical process. But quantum mechanics doesn't describe it. Does quantum mechanics need to be modified or extended to include it?

Von Neumann suggested that the experimenter's awareness of the outcome of a quantum measurement causes the superposition of outcome states to collapse. Other prominent physicists, including Eugene Wigner, embraced this view. Wigner interpreted a system's wavefunction as representing the physicist's knowledge about the system: "Given any object, all the possible knowledge concerning that object can be given as its wave function."[5]

> [T]he impression which one gains at an interaction [with a quantum system] may, and in general does, modify the probabilities with which one gains the various possible impressions at later interactions. In other words, the impression which one gains at an interaction, called also the result of an observation, modifies the wave function of the system. The modified wave function is, furthermore, in general unpredictable before the impression gained at the interaction has entered our consciousness: it is the entering of an impression into our consciousness which alters the wave function because it modifies our appraisal of the probabilities for different impressions which we expect to receive in the future.[6]

In an essay entitled "The Copenhagen Interpretation of Quantum Theory" Werner Heisenberg (1958) argued that the indeterminacy of quantum measurement outcomes arises from "our incomplete knowledge of the world." Just before a measurement begins, the

---

5   Wigner later changed his mind.
6   Wigner, Eugene, "Remarks on the Mind-Body Question" in *Symmetries and Reflections*, (Cambridge, MA. MIT Press, 1970) p. 173

measured object is in a definite quantum state. Then, in Heisenberg's account, it interacts with a macroscopic measuring device.

> This influence introduces a new element of uncertainty, since the measuring device is necessarily described in the terms of classical physics; such a description contains all the uncertainties concerning the microscopic structure of the device which we know from [statistical] thermodynamics, and since the device is connected with the rest of the world, it contains in fact the uncertainties of the microscopic structure of the whole world. These uncertainties may be called objective in so far as they are simply a consequence of the description in the terms of classical physics and do not depend on any observer. They may be called subjective in so far as they refer to our incomplete knowledge of the world. ... It is for this reason that the result of the [measurement] cannot generally be predicted with certainty; what can be predicted is the probability of a certain result ...[7]

Physicists Christopher Fuchs and Rüdiger Schack have proposed a related interpretation of quantum theory, which they call Quantum Bayesianism or QBism. (Bayesian statistics interprets probabilities as degrees of belief. Its methods enable one to update his or her degree of belief in a hypothesis in the light of new evidence.) It rests in part on an earlier paper they co-authored with Carlton Caves. Physicist N. David Mermin, a prominent supporter of this interpretation, has written:

> QBism attributes the muddle at the foundations of quantum mechanics to our unacknowledged removal of the scientist from the science.[8]

Much of this muddle is associated with the 'wavefunction' that quantum mechanics assigns to a physical system. This irritatingly uninformative term reveals the lack of clarity present in the field from its very beginning in 1925. People argue to this day about whether wavefunctions are real entities, like stones or ripples on a pond, or mathematical abstractions that help us to organize our thinking, like the calculus of probabilities.

Fuchs and Schack adopt the latter view. They take a wavefunction to be associated with a physical system by an agent – me, for example,

7   Heisenberg, W. *Physics and Philosophy: The Revolution in Modern Science*, (London: George Allen & Unwin, 1958.)
8   Mermin, N. David, *Nature*, 26 March 2014, "QBism puts the scientist back into science"

based on my past experience. I use the wavefunction, following rules laid down by quantum mechanics, to calculate the likelihood of what I might experience next, should I choose to probe further. Depending on what I then perceive, I can update the wavefunction on the basis of that experience, allowing me to better assess my subsequent expectations.[9]

At the opposite pole from QBism are emended versions of Schrödinger's equation that describe the "collapse of the wavefunction" as a real physical process. G.C. Ghirardi, A. Rimini, and T. Weber (1986) have proposed one such theory, and Philip Pearle (1989) has described a modified version of it. Both versions add to Schrödinger's equation a stochastic term that causes a superposition of quantum states to collapse onto one of its components. The stochastic term represents an interaction between a nominally undisturbed system and a purpose-built and otherwise unobserved space-filling stochastic field, reminiscent of the ether of pre-relativity physics. These theories contain adjustable parameters whose values are chosen to ensure that superpositions representing possible states of atoms and molecules do not collapse within observable periods of time whereas superpositions representing impossible states – superpositions whose components are separated by macroscopic distances – do not persist long enough to be detected experimentally.

Still other attempts to explain why measurements have definite but unpredictable outcomes take as their starting point the remark that under typical experimental conditions *no* bounded system (including the supersystems considered in decoherence theories) can be idealized as an undisturbed system in a definite quantum state. In 1957 Hugh Everett III, then a doctoral student of John A. Wheeler, suggested a radical answer to the question: Why do quantum measurements have definite but unpredictable outcomes? He postulated that quantum mechanics is universally valid: it applies to arbitrarily large undisturbed physical systems. But only the universe is an undisturbed physical system. So, Everett reasoned, we should identify the supersystem consisting of the measured system, the measuring device, and the environment with the universe. A quantum measurement then causes the premeasurement state of the universe to evolve into a superposition of universe-states in each of which the

---

9   Caves, C. M., Fuchs, C. A., & Schack, R. (2002). "Quantum probabilities as Bayesian probabilities." *Physical Review* A, 65(2), 022305.

measured system is in an eigenstate of the measured observable and
the rest of the universe is in a correlated quantum state. Everett now
postulated that the universe-states that occur in this superposition
represent states of distinct but equally real universes in each of which
a particular measurement outcome has occurred. A measurement
causes the universe in which it occurred to split into many – sometimes infinitely many – replicas, differing only in the outcome of the
measurement that caused the splitting.

There is an extensive literature about what this account might (or
could) mean, and how (or whether) it can be reconciled with the rest
of physics, including general relativity and the physics of the everyday world of medium-sized objects. As far as I know, these issues
haven't been resolved.

The theories of quantum measurement I've been describing
assume either that some bounded part of the universe can be idealized as being in a definite quantum state or that the universe itself
is in a definite quantum state. None of these theories successfully
predicts that an ideal quantum measurement leaves the measuring
apparatus in a definite pointer state and the measured system in a
corresponding quantum state, an eigenstate of the measured physical quantity. They leave Weinberg's question *"How do probabilities
get into quantum mechanics?"* unanswered. My suggested answer to
this question also answers a question asked in 1908 by the mathematician and theoretical physicist Henri Poincaré in his essay "Chance":
How do probabilities get into classical mechanics?

Poincaré's "Chance"

Poincaré begins his essay by contrasting the "ancients'" view of
chance with the "modern" view:

> To begin with, what is chance? The ancients distinguished between
> the phenomena which seemed to obey harmonious laws, estab-
> lished once for all, and those that they attributed to chance, which
> were those that could not be predicted because they were not sub-
> ject to any law. In each domain the precise laws did not decide
> everything, they only marked the limits within which chance was
> allowed to move. In this conception, the word chance had a pre-
> cise, objective meaning; what was chance for one was also chance
> for the other and even for the gods.

But this conception is not ours. We have become complete deter-
minists, and even those who wish to reserve the right of human
free will at least allow determinism to reign undisputed in the
inorganic world. Every phenomenon, however trifling it be, has
a cause, and a mind infinitely powerful and informed concerning
the laws of nature could have foreseen it from the beginning of the
ages. If a being with such a mind existed, we could play no game of
chance with him; we should always lose.

For him, in fact, the word chance would have no meaning, or
rather there would be no such thing as chance. That there is for
us is only on account of our frailty and our ignorance. And even
without going beyond our frail humanity, what is chance for the
ignorant is no longer chance for the learned. Chance is only the
measure of our ignorance. Fortuitous phenomena are, by definition, those whose laws we are ignorant of.[10]

No sooner does Poincaré arrive at this answer to his opening question than he explains why it won't do. Chance, he argues, must be
"something more than the name we give to our ignorance," because
some chance phenomena, while individually unpredictable, obey
mathematical laws. "Laws of chance" relate the probabilities of measurement outcomes and other macroscopic phenomena to the frequencies of these outcomes in large (or infinite) collections of identical trials. For example, probability theory allows us to calculate the
probability of k heads in N tosses of a coin, for $k = 0, 1, 2, ..., N$, if
we are given the probability p of heads for a single toss. It predicts
that the probability of pN heads approaches 1 as N increases without
limit.

Poincaré now asks, "What are the defining properties of chance
processes whose outcomes are governed by probabilistic laws?" He
considers three examples of such processes.

The first is an idealized cone balanced on its tip. If no force other
than gravity acts on it, the cone remains in this state forever, but
any disturbance, however slight, causes it to topple over. If we try to
balance a real cone on its tip we will fail. The cone will topple in an
unpredictable direction. But if we repeat the experiment many times,
using exactly the same experimental arrangement to fix the cone's
initial position and motion, the direction in which it tips will be
more or less uniformly distributed between 0 and 2π radians (0 and
360 degrees). If the cone tended to topple in a particular direction,

10   Poincaré, H. *Science and Method*, (New York, Dover Publications, 2003), p.64-5.

we would look for the cause of this asymmetry in a corresponding asymmetry of the cone itself or of the experimental arrangement that determines its initial state. In this example, as Poincaré put it. "slight differences in the initial conditions produce very great differences in the final phenomena." Roulette, another of Poincaré's examples, exemplifies the same principle. Small uncontrollable differences in the relative speed of the ball and the roulette wheel produce unpredictable differences in the number of red and black compartments it passes over before coming to rest. But probability theory predicts that the fraction of N spins with the outcome red approaches ½ as N increases.

Poincaré discusses other processes that display extreme sensitivity to small changes in their initial conditions. In the 1880s he had discovered that in some regions of the planetary system, a small change in an asteroid's initial position or velocity causes its orbit to diverge at an exponential rate from the original orbit. Such orbits are predictable in principle but not in practice. They are said to be *chaotic*; and the phenomenon they exemplify is called deterministic chaos. Students of chaos theory have found that mathematically analogous processes occur in a wide variety of complex systems, including the Earth's atmosphere and the biosphere.

The processes Poincaré discusses have unpredictable outcomes not because we don't know the laws that govern them – we do – but because, unlike an "infinitely powerful and informed" mind, we don't know enough about their initial conditions. Yet we correctly predict that the cone topples in a random direction, that in the long run the roulette ball comes to rest in a red compartment as often as it does in a black compartment, and that in the planetary system small bodies in chaotic orbits are randomly distributed along the ecliptic (the great circle in which the plane of Earth's orbit intersects the celestial sphere). Why do these statistical predictions succeed?

Consider roulette. If we knew the initial speeds of the ball and the wheel we could predict the color of the compartment in which it came to rest. But the initial speeds have ranges of possible values. Define subranges of these ranges so that initial relative speeds in the same subranges correspond to the same outcome. If the number of these subranges is large enough and if the initial relative speeds are smoothly – not necessarily uniformly – distributed over the whole range of possible initial speeds, then the probabilities of the outcomes red and black will be nearly equal, because any smooth curve has nearly the same height at nearly equal horizontal coordinates. Thus, Poincaré concludes, red and black occur equally frequently in a long series of identical spins of the wheel because two conditions are satisfied: (a) small changes in a relevant initial condition produce large differences in the outcome (i.e., the number of same-outcome subranges in the range of initial conditions is very large); and (b) the distribution of initial conditions is smooth, so that contiguous subranges have (nearly) equal probabilities of being realized. Poincaré's other examples admit a similar analysis.

Condition (a) defines a class of chance processes that have statistically predictable outcomes. Poincaré's remaining task was to explain why, in processes that satisfy this condition, the curve that represents the distribution of the relevant initial conditions is smooth.

Poincaré argued that the distribution of initial conditions of chance processes that have statistically predictable outcomes is the "outcome of a long previous history," during which irregularities on the smallest scales have been smoothed out by "complex causes" working over a long period of time.

Poincaré doesn't appeal to the second law of thermodynamics to justify this claim. (As I argued earlier, such an appeal would be illegitimate because the Second Law doesn't apply to the physical universe.) He does, though, connect the absence of irregularities on very small scales to the temporal asymmetry of macroscopic processes – the fact that macroscopic processes such as heat flow, unlike molecular motions and collisions, are governed by laws that distinguish between the direction of the past and the direction of the future. He discusses in some detail how different the course of events would look to a being traveling backward in time and concludes: "Lengthy explications [of such differences] perhaps would aid in the better comprehension of the irreversibility of the universe." In the end, he leaves the smoothness of the curve that represents the probability distribution of the cone's initial tilt unexplained.

Poincaré's Cone in Light of the Strong Cosmological Principle

Poincaré assumes that the initial positions and velocities of the cone's axis, the roulette ball, and the asteroid have definite values. In other words, these systems are in definite (classical) microstates. Since we don't know the precise values of the physical quantities that characterize classical microstates we assign probabilities to them.

A probability distribution of initial microstates defines a (classical) macrostate. A macrostate is the result of an experimental or observational protocol. An experiment or observation gives us information about the macrostate of the measured or observed system.

Newton's laws of motion connect initial microstates to final microstates. Hence they connect the probability distribution that characterizes a system's initial *macrostate* to the set of its possible final *macrostates*. What Poincaré couldn't explain was why we can successfully represent the probability distributions of microstates that characterize the initial macrostates of the cone, the roulette ball, and the asteroid by smooth curves.

In a universe in which the strong cosmological principle holds, macroscopic systems don't in general have definite initial microstates. Instead a macroscopic system's initial macrostate is characterized by a probability distribution of microstates fashioned by the system's history. The most recent episode of such a history describes the act of measurement or observation, and the probability distribution it creates typically contains particular kinds of information but not others. An experimental arrangement designed to put the cone into its initial state imposes limits on the cone's initial tilt and initial angular velocity but creates little or no additional information. The device that sets the roulette wheel and the ball in motion determines their ranges of initial speeds but nothing more. And plausible scenarios for the origin of the solar system don't specify the initial distribution of the positions and velocities of each of the myriad small bodies it contains in enough detail and with enough precision to produce a non-uniform distribution of their present celestial longitudes. If the strong cosmological principle holds, *the probability distributions that characterize the initial macrostates of the macroscopic systems in Poincaré's examples are represented by smooth curves because in each case the system's history didn't create the kinds of information that would produce deviations from smoothness.*

Instead of imagining an infinite series of identical trials whose initial conditions are defined by a probability distribution of (initial) microstates, we can imagine an *experimental ensemble* – an infinite collection of identical cones uniformly distributed in space, each in a definite microstate. We stipulate that the fraction of cones in every range of microstates equals the fraction of identical trials in which the cone's initial microstate lies in that range. That fraction is the probability assigned to that range by the probability distribution that characterizes the cone's initial macrostate. Such a collection evolves into one in which the fraction of cones in every range of microstates equals the fraction of trial outcomes in which the cone's final microstate lies in that range.

This mode of description extends to experimental systems the indeterminacy in position that the strong cosmological principle attributes to naturally occurring systems. It says that physics doesn't describe particular systems but only uniformly dispersed and isotropically oriented systems defined by their histories.

The strong cosmological principle and the assumption of primordial randomness also explain what Poincaré called "the irreversibility of the universe." In a history of the physical universe based on these initial conditions, randomness is the primordial condition. Subsequently, cosmogonic processes create information associated with the probability distributions that characterize galaxies, stars, planets, and other self-gravitating systems. Some of this information decays in processes governed by the second law of thermodynamics, and thereby fuels the creation of new kinds of information. For example, the burning of hydrogen to helium in the Sun's core creates the sunlight that supports life on Earth. Do these inferences about the strong cosmological principle still hold in a universe whose microstructure is governed by quantum mechanics rather classical physics? Can the strong cosmological principle resolve the problem of time's arrow and the measurement problem of quantum mechanics?

Quantum Indeterminacy and Quantum Measurement

To extend Poincaré's account to macroscopic systems whose microstates are governed by quantum physics we need a way of linking measurement outcomes, which give us information about a mea-

suring device's classical macrostate, to the microstates of the measured system, described in the language of quantum physics. Dirac's
general assumption and its generalization[11] supply such a link.

Let s denote a microstate of a system S, and let O denote a property
S. Suppose S is in the microstate s.

> Dirac's general assumption equates a real number,
> denoted by $\langle s|O|s \rangle$, that depends on the state s and the
> observable O to the average result of a large number of
> measurements of O when the system is in state s.

Let α denote a macrostate of the macroscopic system S. Gibbs's
statistical mechanics characterizes such macrostates by probability
distributions of the system's microstates. Let s denote a microstate
of S, and let $p_\alpha(s)$ denote the probability assigned to that microstate
by the probability distribution that characterizes the macrostate α.
Dirac's general assumption implies that the weighted average of the
quantities $\langle s|O|s \rangle$ – the sum of products $\langle p_\alpha(s) \; s|O|s \rangle$ – equals the
average result of a large number of measurements of O when the
system is in the macrostate α:

> The average of the quantities $s|O|s$, weighted by the
> probability that the microstate s is in the macrostate α,
> equals the average result of a large number of measure-
> ments of O when the system is in the macrostate α.   (1)

In a description of measurement that comports with the strong
cosmological principle we interpret the initial conditions that characterize a macroscopic system S as characterizing an experimental
ensemble of systems, and the preceding rule becomes:

> The weighted average in rule (1) equals the average value
> of the macroscopic counterpart of O in an experimental
> ensemble of systems in the macrostate α.   (2)

We can now describe quantum measurements in a way that differs
slightly but significantly from von Neumann's account as emended
by decoherence theory. Decoherence also plays an essential role in
the new account: it randomizes the relative phases of wavefunctions of microstates of the combined system that belong to different

11   Dirac, P.A.M., *The Principles*, pp. 132-133.

pointer states. But the new account, unlike von Neumann's account as emended by decoherence theory, doesn't assume that the combined system plus a random section of its environment is in a definite quantum microstate. Instead it assumes that the combined system plus a random section of its environment is in the superposition of microstates that, in the quantum version of Gibbs's statistical mechanics, represents a macrostate of the combined system. A straightforward calculation then shows that the experimental ensemble that represents the premeasurement state of the combined system rapidly evolves into an experimental ensemble of postmeasurement states in each of which the measured system is in an eigenstate of the measured observable and the measuring apparatus is in the corresponding pointer state.

Does this account of quantum measurement explain how probability gets into quantum mechanics? Does it solve the measurement problem?

One can argue that it does *not* explain how probability gets into quantum mechanics, because it rests on the strong cosmological principle, which *posits* that a complete description of the physical world specifies only probabilities of physical states and physical quantities. Posits are not explanations.

Yet the strong cosmological principle doesn't have an *ad hoc* character. As I mentioned earlier, it defines the simplest possible model of the universe that is consistent with observations of the cosmic microwave background and of the spatial distribution and line-of-sight velocities of galaxies. It also accounts for the fact that reference frames in which Newton's and Maxwell's laws as well as their special-relativistic generalizations hold are unaccelerated relative to a frame defined by the cosmic microwave background and the distribution and motions of distant galaxies. Equally importantly, the strong cosmological principle supplies a single, objective interpretation of probability as it occurs in quantum physics, in statistical mechanics, and in cosmology. Does the preceding account of measurement solve the measurement problem? One can argue that it doesn't.

Dirac's general assumption equates the number denoted by $\langle s|O|s \rangle$ to the average result of a large number of measurements of

O. Von Neumann's account of measurement was intended to replace this assumption by a theory of measurement. It failed to do so, even when emended by decoherence theory. The preceding account of measurement unlike the emended von Neumann account, successfully predicts that quantum measurements have definite outcomes. *But it relies on Dirac's general assumption* (which underlies the quantum version of Gibbs's statistical mechanics). Doesn't this mean that it leaves the measurement problem unresolved?

Yes and no. Yes if you assume that there are as-yet undiscovered mathematical laws of which the laws of quantum physics and classical physics (including general relativity) are limiting cases. If such laws existed, Dirac's general assumption wouldn't be an assumption; it would be a theorem. In their present forms, however, quantum physics and classical physics have unbridgeable differences. Quantum superpositions have no classical counterpart; and quantum physics can't describe curved spacetime. At the same time, special relativity is a pillar of quantum mechanics. The last point is worth a more detailed discussion.

Special Relativity and Quantum Mechanics

Einstein based his special theory of relativity (1905) on two assumptions: a) All inertial reference frames – frames in which Newton's laws of motion and Maxwell's laws of electricity and magnetism hold – enjoy equal status; b) The speed of light in empty space has the same value in every direction and in every inertial frame. Minkowski accordingly set the speed of light in empty space equal to 1, so that in his new geometry distance intervals and time intervals are always measured in the same unit. He then defined the squared space-time interval between two point-events in spacetime as the squared time interval between the point-events and the squared distance interval (given by Pythagoras's theorem as the sum of their squared position-coordinate differences); and he postulated that this quantity has the same value in every unaccelerated reference frame. Finally, he stipulated that the algebraic statements that express physical laws should take the same form in all unaccelerated, or inertial, reference frames.

These postulates greatly extended the domains of mechanics and electromagnetism. The laws of pre-relativity physics became limiting cases of relativistic laws, valid only for particles traveling with speeds much less than the speed of light. And experiments invariably confirmed the predictions of theories that comported with the principles of special relativity. Nevertheless, the earliest versions of quantum mechanics, Werner Heisenberg's "matrix mechanics" of 1925 and Erwin Schrödinger's "wave mechanics" of 1926, didn't comport with these principles: like Newton's laws of motion, they assumed that space is Euclidean and that the time interval between any two moments has the same value in all allowed coordinate systems.

Schrödinger's wave equation is the quantum counterpart of Newton's equation of motion for an electron in an external electric field. Schrödinger and other founders of quantum theory tried to formulate a relativistic generalization of the wave equation – an equation that would have the same form in all unaccelerated, or inertial, reference frames. But they ran into a formidable mathematical obstacle. Schrödinger's equation is "linear": any sum of numerical multiples of two (or more) solutions of the equation that satisfy the same boundary conditions) is likewise a solution. There were compelling reasons to require the equation's relativistic generalization to share this property. But that requirement seemed impossible to satisfy.

In 1928 Paul Dirac discovered a way around this difficulty. He devised a wave equation whose coefficients were not numbers but 4×4 matrices – rectangular arrays of (real or complex) numbers that can be added to and multiplied by one another according to certain rules. These matrix-coefficients weren't arbitrary: they were determined by the requirement that the relativistic wave equation be consistent with special relativity's version of the law of energy conservation. Three years earlier, Heisenberg had used matrices to represent an electron's position coordinates and momentum components, but no physical law with matrices as coefficients had previously been suggested.

Schrödinger's equation has a single solution that satisfies given boundary conditions. It represents the quantum state of a point-like charged particle. Dirac's relativistic wave equation has not one but four solutions.

One pair of these solutions describes electrons with an extra, internal degree of freedom. Dirac proved that the extra degree of freedom shows up in two ways: as an intrinsic angular momentum, or spin;

and as an intrinsic magnetic moment. A classical point charge has neither an intrinsic angular momentum nor an intrinsic magnetic moment. Dirac showed that his relativistic wave equation predicts that electrons have both. It also predicts that the spin has two possible values, $h/4\pi$ and $-h/4\pi$, where h is Planck's constant. And it predicts the ratio between the intrinsic magnetic moment and the spin. The predicted ratio has twice the value classical electromagnetic theory predicts for a spinning charged sphere. Experiments confirmed both predictions with high accuracy. For example, Dirac's theory predicts that each energy level of an electron in an external magnetic field splits into a pair of levels, one with the electron's magnetic moment parallel to the magnetic field line at its position, the other antiparallel; and it predicts the magnitude of the splitting.

Dirac's wave equation also made new predictions about the energy spectrum of hydrogen. Schrödinger's equation predicts that the electron in a hydrogen atom has a discrete set of negative energy states. The energy of the nth state, for $n = 1, 2, 3, ...$, is inversely proportional to n^2. Schrödinger's equation also predicts the (negative) constant of proportionality. Both predictions agree closely with experiment. Dirac's relativistic theory predicts that the energy levels have a fine structure that depends on a second quantum number, j, which also takes integer values. This prediction matched precise experimental measurements.

The second pair of solutions to Dirac's relativistic wave equation was harder to interpret than the first pair. It seems to describe electrons in impossible states of negative total energy (including the rest energy mc^2). Dirac eventually concluded that the negative-energy solutions represent a previously unknown particle with the same mass and spin as the electron and an equal but opposite charge. A short time later, experimental physicists discovered a particle with precisely these characteristics, the positron.

Despite its successes, Dirac's relativistic theory of the electron raised important technical problems. "[T]hese problems were all eventually to be solved (or at least clarified) through the development of quantum field theory."[12] At the same time, quantum field theory reinforced the link between quantum mechanics and special rela-

---

12   Weinberg, Steven, *The Quantum Theory of Fields*, (Cambridge University Press, 1995,) Volume 1, p. 14.

tivity. In the preface to his magisterial two-volume *The Quantum Theory of Fields*, Steven Weinberg writes: "The point of view of this book is that quantum field theory is the way it is because ... it is the only way to reconcile the principles of quantum mechanics ... with those of special relativity."[13]

Special relativity, in turn, is a local approximation to general relativity, Einstein's unified theory of spacetime and gravitation. In Minkowskian spacetime the squared spacetime interval between two events equals the squared time interval between the events minus the squared distance interval, which in turn equals the sum of the three squared coordinate intervals. Gauss showed that the geometry of a smoothly curved surface could be derived from a formula that expresses the squared distance between two neighboring points on the surface as a sum of multiples of the squares of the coordinate intervals in a system of curvilinear coordinates, in which the multiples vary smoothly with position. Riemann, Gauss's pupil, extended his theory to curved spaces with any number of dimensions. Riemann's theory expresses the squared distance between two neighboring points as a sum of multiples of the squares of the coordinate intervals in an n-dimensional system of curvilinear coordinates; the multiples vary smoothly with position. Finally, Einstein's theory expresses the squared distance between two neighboring events in spacetime as the difference between a multiple of the squared time interval between the events and a sum of multiples of the squared space intervals in a four-dimensional system of curvilinear coordinates; each of the four multiples is a smooth function of the four coordinates. Einstein's field equations relate these multiples and their first and second derivatives to the spacetime distribution and flow of mass/energy. *Quantum mechanics presupposes that the (undisturbed) systems it describes are embedded in the spacetime Einstein's general theory of relativity describes.*

For decades physicists have tried to construct a mathematical structure that would contain general relativity and quantum mechanics as limiting cases. Yet, as we've seen, quantum mechanics needs general relativity; and general relativity doesn't explain the microstructure of the physical world. The preceding discussion shows that in a universe that comports with the strong cosmological principle, quantum

---

13   ibid. p. xxi

mechanics and general relativity fit smoothly together, linked by the quantum version of Gibbs's statistical mechanics and the ensemble version (2) of Dirac's general assumption (1). In this scheme chance arises from the postulated existence of space-time coordinate systems relative to which a complete description of the physical world privileges no position or direction, and manifests itself in quantum indeterminacy, in the probabilities that relate microstates (governed by quantum mechanics) and macrostates (governed by classical physics), and in the largely indeterminate character of the initial conditions that define macroscopic objects and processes.

VII

Living World

Physicalism rests on the assumption that the physical quantities that occur in classical, or non-quantum physical theories, have definite values even when we don't or can't know them precisely. It implies a modified form of determinism that allows for quantum indeterminacy. As an alternative to physicalism I've proposed a pair of cosmological hypotheses: the strong cosmological principle and the assumption of primordial randomness. I've argued that these hypotheses define a framework for physics that supplies physical interpretations of chance in quantum mechanics and statistical mechanics. Now I want to argue that this framework allows us to view biology as an autonomous science – one that, though based on strongly confirmed physical laws, doesn't reduce to physics. In particular the new framework accommodates consciousness and, more generally, the inner lives of living organisms.

Biological Complexity

Although the same strongly confirmed physical laws govern physical and biological processes, Ernst Mayr has argued that living systems exhibit kinds of complexity that have no counterpart in the physical world. The organization of living systems

> endows them with the capacity to respond to external stimuli, to bind or release energy (metabolism), to grow, to differentiate, and to replicate. Biological systems have the further remarkable prop-

erty that they are open systems, which maintain a stead-state balance in spite of much input and output. This homeostasis is made possible by elaborate feedback mechanisms, unknown in their precision in any inanimate system.[1]

Mayr immediately points out that "complexity in and of itself is not a fundamental difference between organic and inorganic systems." But organic complexity has "extraordinary properties not found in inert matter"[2]:

> The complexity of living systems exists at every hierarchical level, from the nucleus to the cell, to any organ system (kidney, liver, brain), to the individual, to the species, the ecosystem, the society. The hierarchical structure within an organism arises from the fact that the entities at one level are compounded into entities at the next higher level – cells into tissues, tissues into organs, and organs into functional systems.[3]

Moreover, "systems at each hierarchical level have two properties. They act as wholes ... and their characteristics cannot be deduced from the most complete knowledge of the components taken separately ..."[4] In other words, systems at each level are characterized by emergent properties – properties that evince novel and unpredictable kinds of order.

By contrast, the kinds of order that physicists study are predictable, at least in principle. As an example consider the orderly structure of hemoglobin, the molecule in the blood of vertebrates that carries oxygen from the lungs to the rest of the body. Quantum mechanics predicts the structure of hemoglobin. It also predicts that the molecule has two distinct conformations with different oxygen-binding affinities. But chemists, as chemists, don't ask how these two conformations came about – why they occur in the blood of vertebrates – or why the active form is found in the lungs but flips into the inactive form in muscle tissues undergoing exertion. Quantum mechanics doesn't address these questions. More generally, it doesn't seek to explain the *functions* and *histories* of specifically biological kinds of complexity – kinds of complexity that emerge in the course of evolution.

---

1   Mayr, Ernst, *Toward a New Philosophy of Biology* (Cambridge, Harvard University Press, 1988), p. 14.
2   ibid.
3   ibid.
4   ibid, p.15.

Chance and Order

Earlier I quoted Mayr on the importance of chance in evolution. Chance, he points out, plays a central role both in genetic variation and natural selection. It is likewise central to our pre-scientific views of individual human experience and human history. Yet in the physicalist world-view only quantum measurements have objectively unpredictable outcomes, and quantum measurements are largely irrelevant to both genetic variation and natural selection. In the physicalist worldview the unpredictability of events that are not quantum measurement reflects our ignorance of relevant initial conditions. In contrast, in a scientific worldview that comports with the strong cosmological principle chance and the assumption of primordial randomness chance is objectively real at all levels of description, from the molecular to the cosmological. Randomness is the raw material from which processes governed by deterministic mathematical laws fashion myriad novel varieties of physical and – especially – biological order

Randomness and Information Have a Hierarchical Structure

In a scientific theory that comports with the strong cosmological principle physical quantities are associated with probability distributions. Such a theory assigns a physical quantity's possible values probabilities and assigns each of these values (or each small range of values) a probability. It interprets these probabilities as relative frequencies in a cosmological ensemble. Following Boltzmann, I defined the *randomness* of a probability distribution as the mean, or probability-weighted average, of the negative reciprocal of the probability, and the probability distribution's *information* as the amount by which the distribution's randomness falls short of its largest possible value – the value that meets given constraints. Claude Shannon showed that the information of a probability distribution is the sum of contributions associated with statistical correlations between pairs, triples, quadruples, and so on of the events to which the distribution assigns probabilities.

If the assumption of primordial randomness holds, the probability distributions that characterized the universe at the earliest times when our present strongly confirmed physical laws held were maximally random; they contained no information. As the universe expanded, physical processes governed by quantum mechanics and by the laws of general relativity created information in the form of multiparticle correlations. In particular, gravitational instability – the tendency of relatively dense regions to expand more slowly than less dense regions – caused the initially uniform distribution to become progressively more clumpy.[5] This process creates information on very large scales, leaving randomness on molecular and small macroscopic scales untapped.

Once self-gravitating systems have begun to form, gravitational collapse – the tendency of relatively dense regions to expand more slowly than less dense regions –creates spatial inequalities of mass density and temperature. Thus gravitational collapse diminishes randomness and creates information. It is the ultimate source of the information that, in accordance with Clausius's law of entropy non-increase, fueled the origin of, and continues to sustain, life on Earth.

In experimental physics and chemistry, the experimenter uses a local source of information to create the information that characterizes an experimental setup. As Bohr emphasized, experimental setups must be capable of being described in the language of classical physics. So descriptions of experimental setups don't contain quantum-level information. Of course, experimental *predictions* can and do involve quantum mechanics.

(Think of the Stern-Gerlach experiment.)

Biophysicists and biochemists use the experimental and theoretical methods of physics and chemistry to study specific biological phenomena. Their efforts leave little room for doubt if any that biological structures and processes are also physical structures and processes, governed by the same strongly confirmed physical laws. Yet the theory of biological evolution, within which accounts of specific biological structures and processes are necessarily embedded (recall Dobzhansky's dictum that nothing in biology makes sense except in the light of evolution), can't be construed as a physical theory. Physical theories make testable predictions about the outcomes of experiments or observations based on assumptions about initial

---

5   For a detailed account of one version of this process, see David Layzer, *Cosmogenesis: The Growth of Order in the Universe*, (Oxford, Oxford University Press, 1990)

conditions. In contrast, evolution gives rise to novel and unpredictable kind of order.

How do evolutionary explanations differ from explanations offered by physical theories? Evolutionary theories, like theories in the physical sciences, seek to show that the present state of a system could have evolved from a hypothetical earlier state through processes governed by physical laws. Laplace's theory of the origin of the solar system, for example, sought to explain why the system consists of planets that circle the Sun in the same direction. The theory purported to show that an initially structureless gas cloud governed by Newton's laws would evolve into a system with these features. Analogously, population genetics, the predictive arm of evolutionary theory, successfully models the evolutionary acquisition of relatively simple traits like industrial melanism, "the darkness—of the skin, feathers, or fur—acquired by a population of animals living in an industrial region where the environment is soot-darkened."[6]But population genetics doesn't aspire to predict, for example, that some species of fish will evolve into land-dwelling animals. We don't yet have a sufficiently detailed and reliable description of the evolutionary precursors of land-dwelling animals. And, as Mayr has emphasized, the evolutionary history of land-dwelling animals is punctuated, and to a considerable extent determined by, chance events that have major but unpredictable evolutionary consequences. The story that connects a structureless gas cloud to a model of the solar system can't have an evolutionary counterpart because chance events break the causal chain.

What Is Life?

Although the origin of life is still highly speculative, biologists' understanding of the history of life leaves little room for doubt that all living organisms descended from small populations of membrane--enclosed, approximately self-replicating collections of molecules. What distinguished these primordial organisms from nonliving molecular assemblies of similar composition and complexity?

Unlike nonliving molecular assembles, every living organism, no matter how simple, has interests and *needs* that set it apart from and put it into a particular relationship with its environment. A living

---

6   "Industrial Melanism", *Encyclopedia Britannica*.

organism must regulate its interaction with its surroundings in ways that allow it, among other things, to:

- replicate itself more or less accurately;
- maintain processes that allow it to import building materials and high-grade (i.e. low-entropy) information and export waste products and low-grade information;
- maintain close-to-optimal relative chemical abundances
- (homeostasis); protect itself against predators.
- promote the wellbeing and reproductive success of other, not necessarily related, individuals and groups.

These and other interests and needs not only distinguish organisms from nonliving systems. They also distinguish living organisms from one another. Every living organism has a unique set of interests and needs as well as a unique history.

The observation that sensitivity to external stimuli, sentience, and consciousness are ubiquitous in the living world suggests that *the emergence of enclosed, approximately self-replicating molecular assemblies coincided with the emergence of something like a point of view – a rudimentary version of subjectivity.* Inner lives enable living organisms to respond in unified and creative ways to the challenges and opportunities presented by the external world and to exploit the opportunities offered by the ubiquity of chance in the macroscopic world.

This hypothesis may seem daring but it's hard to avoid. Because evolution is a fact and consciousness and sentience are biological attributes, they must be the product of evolutionary processes, like bilateralism or the structure of hemoglobin. Since the most essential feature of consciousness is its subjective character, we can hardly avoid assuming that subjectivity came into being with our earliest ancestors, the first enclosed, approximately self-regulating molecular aggregates with interests and needs that include those mentioned above. How primitive forms of subjectivity evolved into more complex forms then becomes a scientific problem, like the origin of feathers.

Why then do many contemporary neuroscientists and philosophers deny the reality of consciousness as a feature of physical reality distinct from the neural processes that underpin it? The thesis that elementary particles and photons are the ultimate and only constituents of physical reality is neither required nor implied by any strongly confirmed physical law. One can deny it without accepting the strong cosmological principle and the assumption of primordial randomness. But someone who does accept these assumptions has a strong motive to reject physicalism because it implies (and physicalism denies) that chance, randomness, and order, along with the novel kinds of order that biological evolution fashions from the raw material of randomness are as real and objective as mass and electric charge.

Creativity

In a paper entitled "Evolution as a Creative Process" Theodosius Dobzhansky wrote: "A living organism resembles a work of art, and the evolutionary process resembles the creation of a work of art."[7] The creation of a work of art involves analogues of random genetic variation and natural selection. But I think it involves something more. Creative processes not only rely on macroscopic randomness, a feature of the physical world that is a consequence of the strong cosmological principle and the assumption of primordial randomness. They also exploit this feature. Consider the following example of the creative process, supplied by a member of my family.

My brother Bob is a lyric poet as well as a neurologist. I asked him to write down an account of how he writes a poem. This is what he wrote:

HOW I WRITE A POEM
by Robert Layzer

A new poem often starts with a phrase that evokes some interest or feeling, or points to a subject that might be interesting or moving. Sometimes the phrase describes something I saw or heard in the environment, or a painting that I am attached to; or it could be a fragment of a thought or a feeling.

---

7   Dobzhansky, T., *Proc. Ninth Int. Congress Genetics*, 1954, pp. 435-449.

Rarely, the poem may emerge and be completed within a day. More often nothing much happens for a while-hours or days-while I repeat the phrase silently or aloud now and then. Then, out of nowhere, new phrases and sentences start to cluster around the original idea, like crystals coming out of solution. I begin to see a theme, and that suggests more words to enlarge the theme. My conscious mind becomes more involved as I shape and plan the structure and content. Pruning and revising may take place as the poem develops, or after it is nearly finished. This is a much more conscious process; now I am reading the poem in the context of the history of literature, as if I were a critic.

During the composition, I try to listen for unexpected words or ideas, and to the internal assonances and other harmonies that tell a story of their own. In fact, the sound of the poem may turn out to be the main carrier of meaning, and if the poem is successful I may not understand the real "message" for some time afterward. But if I'm successful, the ending often gives a sense of a question answered.

The creative process described in this account generates candidates for selection by a process that is both random and constrained: only candidates in certain broad categories, which themselves depend on the work being created, present themselves for selection. The goal of a creative process isn't known beforehand; the process isn't teleological. But once the process has reached its endpoint, its product can be seen to satisfy "fitness criteria" that were already in place; it "gives a sense of a question answered"; "everything comes together, and you say 'that's it.'"   The sustained exploratory effort in artistic creation has a counterpart in Bergson's philosophical account of evolution: the *élan vital*. It also has counterparts in modern accounts of biological evolution. Notably, biologists John Gerhart and Marc Kirschner have proposed a theory of phenotypic variation in evolution – a theory of how differences between anatomical, physiological, and behavioral traits of organisms have evolved – that differs radically from older accounts, in which phenotypic variation and the divergence of genetic lineages result mainly from the action of natural selection on uncorrelated variations of individual genes:

Most anatomical and physiological traits that have evolved since the Cambrian are, we propose, the result of *regulatory* changes in the usage of various members of a large set of conserved core components that function in development and physiology. Genetic change of the DNA sequences for regulatory elements of DNA, RNAs, and proteins leads to heritable regulatory change, which specifies new combinations of core components, operating in new amounts and states at new times and places in the animal. These new configurations of components comprise new traits.[8]

As I mentioned, artistic creation is an exploratory process. An artist searches for and eventually finds targets in a large "space" of imagined structures. Kirschner and Gerhard have argued that evolution is an exploratory process in this sense:

As the name [Exploratory Processes] implies, some conserved core processes appear to search and find targets in large spaces or molecular populations. Specific connections are eventually made between the source and target. These processes display great robustness and adaptability and, we think, have been very important in the evolution of complex animal anatomy and physiology. Examples include the formation of microtubule structures, the connecting of axons and target organs in development, synapse elimination, muscle patterning, vasculogenesis, vertebrate adaptive immunity, and even behavioral strategies like ant foraging. All are based on physiological variation and selection. [9]

Kirschner and Gerhart have given a more extended (and highly readable) account of their view of evolution in *The Plausibility of Life*.[10]

In 1980 I discussed the hypothesis that evolution is a process of hierarchic construction, involving not just constraints on deleterious mutations but also what Kirschner and Gerhart call "deconstraints."

---

8   Kirschner, Marc W. and Gerhart, John C., *Proc. National Academy of Sciences* 104, supp. 1, May 15, 2007.

9   ibid.

10   Kirschner, Marc W. and Gerhart, John C., illustrated by John Norton, *The Plausibility of Life*, (Yale University Press, 2005).

The only example I then knew about was the adaptive vertebrate immune response. I append the abstract of that paper to this chapter.

Free Will

In his introduction to *The Oxford Handbook of Free Will* philosopher Robert Kane writes:

> [D]ebates about free will in the modern era (since the seventeenth century) have been dominated by two questions, not one – the "Determinist Question": "Is determinism true?" and the "Compatibility Question": "Is free will compatible (or incompatible) with determinism?"[11]

Some scientists and philosophers have argued that free will must be an illusion because genuine freedom – the capacity to influence the course of future event – would be incompatible with determinism; and science, they argue, is deterministic: past conditions and physical laws determine future events. Although quantum mechanics predicts that certain processes – quantum determined measurements – have unpredictable outcomes, such processes don't begin to account for the kind of unpredictability required by a robust conception of human freedom.

If we replace physicalism's assumption that classical physical quantities have definite values by the strong cosmological principle and the assumption of primordial randomness, the answer to Kane's first question – Is determinism true? – is *no*. Initial conditions in biological processes ranging from evolution to individual development and cultural evolution are products of histories punctuated by unpredictable events. And these processes have a creative character. They give rise to novel and unpredictable forms of order.

Since human behavior is a province of biology, we can now replace Kane's second question ("Is free will compatible (or incompatible) with determinism?") by the question "Is free will compatible with biology?" If this book's central argument is correct, the answer is an unambiguous yes.

---

11   *The Oxford Handbook of Free Will*, ed. Robert Kane, Oxford, Oxford University Press, 2002.

Appendix, Chapter VII

Abstract. The present theory offers a unified solution to three closely related evolutionary problems. (1) Why does an evolving population explore only a small fraction of the accessible pathways in genotype space? (2) Conventional ideas about genetic variation suggest that major adaptive shifts, which involve large numbers of separate but functionally related genetic changes, have vanishingly small probabilities of occurrence and require long periods of time. Yet the fossil record indicates that such shifts are neither slow nor uncommon. (3) Studies in comparative morphology and comparative embryology indicate that evolution is a process of hierarchic construction. How is the principle of hierarchic construction related to the basic postulates of evolutionary theory? According to the theory elaborated in this paper, the genomes of both unicellular and multicellular organisms contain two functionally distinct systems of genes: an α system, which encodes a program for the organism's development and which includes regulatory genes and other "modifiers" as well as structural genes; and a β system, which encodes a strategy for adaptive variability and whose elements regulate the rates of genetic recombination and of genic and chromosomal mutations. Corresponding to these two systems are two classes of adaptations: α adaptations, which enhance the fitness of their possessors, and β adaptations, which increase the expectation of fitness in the descendants of their possessors. Sexual reproduction is the most familiar example of a β adaptation. It is noteworthy that the most primitive known form of sexuality, that found among bacteria, is characterized by genetic mechanisms serving to regulate genetic recombination in precisely the manner postulated here for β genes. The β system serves to direct evolutionary flows in genotype space into potentially adaptive channels. It also serves to stabilize and buffer highly adapted genetic structures against the potentially disruptive effects of accidental variations. At the genetic level of description, a major adaptive shift corresponds to the ascent of a fitness peak in a multidimensional subspace of genotype space. The β system serves to focus evolutionary flow in this subspace in the direction of steepest ascent, thereby ensuring rapid and coordinated evolution of the genetic systems involved. (A mathematical framework that should make it possible to construct numerical models of the evolutionary process just described is given by Layzer[12]) The principle of hierarchic construction emerges as a natural consequence of selection-regulated genetic variation. Hierarchic units are defined by their covariability (or costability). The "otherwise inexplicable tendency of organisms to adopt ever more complicated solutions to the problem of remaining alive" (Medawar 1967, pp. 99-100) results from the action (and progressive elaboration) of a genetic system that promotes just those kinds of genetic variation that result in the growth of functionally hierarchic genetic systems. The present theory also throws light on several specific evolutionary problems, including the interpretation of genic polymorphism, the variability of evolutionary rates inferred from the fossil record, the evolution of "pseudoexogenous" and "trivial" adaptation, and the problem of speciation. Applications to the evolution of social behavior are discussed elsewhere.[13]

12   Layzer, David. "A macroscopic approach to population genetics." *Journal of Theoretical Biology* 73.4 (1978): 769-788.
13   Layzer, D., *The American Naturalist*, 115, 6, June 1980.

Afterword

As two researchers who knew well and were deeply influenced by David Layzer's life and work, we are honored to have been able to help his family in assembling, editing, and communicating his final book.

David Layzer believed in and argued for the existence of true novelty, creativity, and freedom in the physical world, and against a worldview of *physicalism* that he saw as regarding all of these as convenient fictions. In that worldview everything we see around us today was "already contained" within the beginning state of the Universe, bound to that early state by an inevitable and barren unfolding via mathematical physical laws.

In this work and others he argued that the flaw in this physicalist view is *not* in the mathematical laws of nature that partly underly it, but rather in the historical and cosmological context in which those laws operate. His postulated Strong Cosmological Principle implies that chance is inevitable: a fully-detailed description of the Universe simply does not exist – or perhaps more accurately, the most-detailed description is *statistical* in the sense that no matter how much data is specified about some region of the Universe, there will be predictive questions with only probabilistic answers. This applies at all times, and only statistical statements about the current universe can possibly follow from its early state. Importantly, while it is widely believed that quantum mechanics implies (some sense of) objective chance, Layzer saw the objective chance inherent in the Strong Cosmological Principle to be the *basis* rather than the *result* of quantum uncertainty, forming a completely original way of viewing the quantum measurement problem that has been taken up by other researchers including one of us (AA).

If the forms and structures inhabiting the current Universe were not (implicitly) present at its beginning, how did this information arise? In this question Layzer has provided insight after insight. He appears to be among the first who crisply connected non-equilibrium processes in cosmology to the formation of chemical and other order – which is now a somewhat textbook view. But he also pointed out how this can be viewed as a competition between entropy generation in a system versus the increase in the maximal possible entropy of that system; an expanding universe provides a widening gap between the two that is cosmological order. This insight has still not been fully concretized in thinking about the statistical mechanics of the Universe, though work in that direction is happening, including by one of us (AA). The cosmic store of order is the grist for the creation of higher levels of order, through astrophysical structure formation, chemical evolution, biological evolution and then neural systems; these combine to transform some of that cosmologically-generated reservoir of order into all of the artifacts we value.

How this creative transformation takes place in the evolutionary and neural domains occupied Layzer through much of his later life, but *Why We are Free* provides only a very compact view, so it is worth here bringing forward a bit more material here, pointing to further sources.

For biology, he approvingly quotes Theodosius Dobzhansky:

> "A living organism resembles a work of art, and the evolutionary process resembles the creation of a work of art."[1]

Layzer's insight into biological evolution endorses the role of chance in the creation of random genetic variations, but he sees subsequent natural selection as a non-chance process. He quotes evolutionary biologist Ernst Mayr:.

> Evolutionary change in every generation is a two-step process: the production of genetically new individuals and the selection of the progenitors of the next generation. The important role of chance at the first step, the production of variability is universally acknowledged, but the second step, natural selection, is on the whole viewed rather deterministically: Selection is a non-chance process.[2]

---

1   Dobzhansky, T., *Proc. Ninth Int. Congress Genetics* (1954), pp. 435-449

2   Mayr, E., *Toward a New Philosophy of Biology*. (Cambridge, Harvard University Press, 1988) p. 21

And Layzer writes

Randomness is the raw material from which processes governed by deterministic mathematical laws fashion myriad novel varieties of physical and – especially – biological order.[3]

For Layzer, chance again plays an essential role in freedom of the will, but it does not make our choices and decisions themselves random, any more than biological evolution created random life forms. Just as randomness seeds genetic variation that is then selected for fitness, random chance simply generates multiple alternative possibilities for thoughts and actions.

In this way Layzer draws a parallel between biological evolution, a two-step process as Ernst Mayr described it, and free will, which has also been described as a two-step process by many thinkers since at least William James in the 1880's.[4] One of us (BD) has identified nearly two dozen philosophers and scientists who have proposed or endorsed two-stage models of free will since James.[5] James was also the first thinker to draw the parallel between Darwinian evolution (then a new idea), freedom of the will, and chance.

In an unpublished manuscript, "Naturalizing Libertarian Free Will," Layzer explained...

It entails a picture of the physical universe in which chance prevails in the macroscopic domain (and hence in the world of experience). Because chance plays a key role in the production of genetic variation and in natural selection itself, evolutionary biologists have long advocated such a picture. Chance also plays a key role in other biological processes, including the immune response and visual perception. I argue that reflective choice and deliberation, like these processes and evolution itself, is a creative process mediated by indeterminate macroscopic processes, and that through our choices we help to shape the future.[6]

But Layzer makes clear that the ultimate decision or choice between alternative possibilities is in no way itself random, but an act of self-determination:

To be fully human is to be able to make deliberate choices. Other animals sometimes have, or seem to have, conflicting desires, but we alone are able to reflect on the possible consequences of differ-

---

3   See chapter VII - Chance and Order, p.131.
4   See Doyle, R.O. "Jamesian Free Will," in *William James Studies* (2010) Vol. 5, p. 1
5   See https://informationphilosopher.com/books/scandal/Two-Stage_Models.pdf.
6   See https://informationphilosopher.com/solutions/scientists/layzer/
Naturalizing_Libertarian_Free_Will.doc. (2010) p.2.

ent actions and to choose among them in the light of broader goals and values. Because we have this capacity we can be held responsible for our actions; we can deserve praise and blame, reward and punishment. Values, ethical systems, and legal codes all presuppose freedom of the will...

A decision is free to the extent that it results from deliberation. Absence of coercion isn't enough. Someone who bases an important decision on the toss of a coin seems to be acting less freely than someone who tries to assess its consequences and to evaluate them in light of larger goals, values, and ethical precepts.[7]

This makes clear that in Layzer's view not all of the Universe is free — that is a capability enjoyed by only a very special set of systems — like human minds — far up the hieheirarchy of complexity. But in his view all of the Universe is creative. From the creation of chemical order in the early universe to the creation of biological and mental novelty, new structures are continually coming into being, and allowing others to come into being. This notion extends even to mental creations such as the mathematical laws of physics that we use to describe the Universe itself. Layzer often quoted Einstein's famous observation that scientific theories are "free creations of the human mind." When one of us (AA) once asked him how to reconcile this with the notion that mathematical statements are true (or not) long before being discovered by people, he gave a beautiful response: many, many, many statements in mathematics follow from a given set of axioms. But this set of "all possible true statements" is *information free* — just like a physical system in equilibrium. Human mathematicians and physicists *select* particular ones from that giant set. In doing so, we create information that did not exist before.

Layzer's work was deeply novel itself, and in many ways unconventional. We hope that this volume might help others to discover and build on some of his many insights, continuing to unfold the creative process he began.

Anthony Aguirre, UC Santa Cruz

Bob Doyle, Harvard University

January, 2021

---

7   See https://informationphilosopher.com/solutions/scientists/layzer/Free_Will_As_A_Scientific_Problem.pdf. (2011) p. 31

Bibliography

Aguirre, A. (1999) *Astrophysical Journal*, 521, 17-29.

Aguirre, A., and Tegmark, M. (2011) "Born in an Infinite Universe: a Cosmological Interpretation of Quantum Mechanics." *Phys. Rev. D* 84.10: 105002.

Ayala, F.J. and T. Dobzhansky., *Studies in the Philosophy of Biology*, Berkeley: University of California Press.

Bergmann, P. J. and J. L. Lebowitz (1955) *Phys. Rev.* 99, 578.

Blatt, J. M., (1959) *Prog. Theor. Phys.* 22, 745.

Boltzmann, L. (1895) *Lectures on Gas Theory*, translated by Stephen G. Brush, New York: Dover Publications, 2011.

Cohen, E. G. D. (1962) *Proceedings of the NUFFIC International Summer Course in Science at Nijenrode Castle*, The Netherlands, August, 1961. Amsterdam: North-Holland Press.

Crick, F. (1994) *The Astonishing Hypothesis*, New York: Simon and Schuster.

Dirac, P.A.M. (1967) *The Principles of Quantum Mechanics*, fourth edition (revised), Oxford: Clarendon Press.

Dobzhansky, T. (1954) *Proc. Ninth Int. Congress Genetics*, pp. 435-449.

Dobzhansky, T. (1974) "Chance and Creativity in Evolution" in Ayala and Dobzhansky, p. 329.

Doyle, Robert O. (2010) "Jamesian Free Will," *William James Studies*, vol.5, p1.

Doyle, Bob (2011) *Free Will: The Scandal in Philosophy*, Cambridge: I-Phi Press.

Eddington, A. S. (1958), *The Nature of the Physical World*, Ann Arbor: University of Michigan Press.

Einstein, A. (1952) *Relativity, the Special and the General Theory*, 15th edition, New York: Crown Publishers.

Einstein, A. (1923) "The Foundations of the General Theory of Relativity," in *The Principle of Relativity*, Methuen and Company, reprinted by Dover Publications, 1952.

Einstein, A. (1953) *The Meaning of Relativity*, 5th edition, Princeton: Princeton University Press.

Fermi, E. (1937) *Thermodynamics*, New York: Prentice-Hall.

Gerhart, J. and Kirschner, M. (2007) *Proceedings of the National Academy of Sciences*, May 15, 2007. 104 (suppl 1) 8582-8589.

Gibbs, J. W. (1902) *Elementary Principles in Statistical Mathematics, Developed with Special Reference to the Rational Foundations of Thermodynamics*, New Haven: Yale University Press.

Greenblatt, S. (2009) *The Swerve*, New York: Norton.

Hahn, E. L. (1950) *Phys. Rev.*, 80, 580.

Heisenberg, W. (1958) *Physics and Philosophy: The Revolution in Modern Science*, London: Allen & Unwin.

Home, D. and A. Robinson, (1995) "Einstein and Tagore: Man, nature and mysticism", *Journal of Consciousness Studies* 2.

Hubble, E. (1936) *The Realm of the Nebulae*, Yale University Press.

Kane, R. (2002) *The Oxford Handbook of Free Will*, New York: Oxford University Press.

Kirschner, M. W. and Gerhart, J. C. (2005) *The Plausibility of Life*, New Haven: Yale University Press.

Kirschner, M. W. and Gerhart, J. C. (2007) *Proc. National Academy of Sciences* 104, supp. 1, May 15, 2007.

Laplace, Pierre Simon, (1951) *A Philosophical Essay on Probabilities*, translated into English from the French 6th ed. by Truscott, F.W. and Emory, F.L. New York: Dover Publications.

Layzer, D. (1980) *The American Naturalist*, 115, 6.

Layzer, D. (1990) *Cosmogenesis: The Growth of Order in the Universe*, New York: Oxford University Press

Lebowitz, J. L and P. J. Bergmann (1959) *Annals of Physics* 1, 1.

Mayr, E. (1983) *The American Naturalist* 121: 324-33; reprinted in Mayr (1988).

Mayr, E. (1988) *Toward a New Philosophy of Biology*. Cambridge: Harvard University Press.

Mermin, N. D. (2014) "QBism puts the scientist back into science", *Nature*, 26 March 2014.

Nagel, T. (2012) *Mind and Cosmos: Why the Materialist Neo-Darwinian Conception of Nature Is Almost Certainly False*, Cambridge: Belknap Press.

Nagel, T. (2013) "The Core of 'Mind and Cosmos," *New York Times*, August 18, 2013.

Norton, J. D. (2015) "The Hole Argument", *The Stanford Encyclopedia of Philosophy* (Fall 2015 Edition), Edward N. Zalta (ed.), URL = <https://plato.stanford.edu/archives/fall2015/entries/spacetime-holearg/>

Penrose, R. (1989) *The Emperor's New Mind*. Oxford: Oxford University Press.

Planck, Max. (1945) *Treatise on Thermodynamics*, trans. Alexander Ogg. Third edition translated from the seventh German edition. New York: Dover Publications.

Ridderbos, T. M. and M. L. G. Redhead, (1988) *Foundations of Physics*. 28, 1237.

Schlosshauer, M. (2008) *Decoherence and the Quantum-to-Classical Transition* (corrected 2d printing) Berlin: Springer.

Schlosshauer, M. (2014) "The quantum-to-classical transition and decoherence," *arXiv*:1404.2635v1 [quant.phys].

Schrödinger, E. (1948) *Statistical Thermodynamics*, Cambridge: Cambridge University Press.

Searle, J. R. (2004) *Freedom and Neurobiology*, New York: Columbia University Press

Shannon, C.E. (1948) *The Bell System Technical Journal*, Vol. 27, pp. 379–423, 623–656, July, October, 1948.

Thomson, W. (1848) *Philosophical Magazine*, October, 1848.

Van Kampen, N. G. (2007) *Stochastic Processes in Physics and Chemistry*, 3d edition. Amsterdam: Elsevier.

Wegner, D.M. (2002) *The Illusion of Conscious Will*. Cambridge: MIT Press.

Weinberg, S. (1972) *Gravitation and Cosmology*. New York: Wiley.

Weinberg, S. (2017) *New York Review of Books*, January 19, 2017.

Weinberg, S. (1995) *The Quantum Theory of Fields*, Volume 1, New York: Cambridge University Press.

Wigner, E. (1970) "Remarks on the Mind-Body Question" in *Symmetries and Reflections*, Cambridge: MIT Press.

Wilson, E.O. (1998) *Consilience*. New York: Knopf.

Index

www.ingramcontent.com/pod-product-compliance
Lightning Source LLC
Chambersburg PA
CBHW070805100426
42742CB00012B/2257